Science, Technology and Gover

Science, Technology and the International Political Economy

Series Editor: John de la Mothe

The upheavals of the international political economy during recent decades have fundamentally altered the relationships between firms and states, citizenship and management, social institutions and economic growth. The changing pace of competition, firm performance and geo-economics is shifting the pressures on public policy and corporate strategy alike. As a result, our conceptual frameworks for analysing key events, emerging trends and driving forces are being challenged. As unclear as the future is, what remains certain is that science, technology and innovation will occupy a central place. By looking at a wide array of issues – ranging from security and foreign affairs, the environment, international institutions, corporate strategy and regional development to research policy, innovation gaps, intellectual property, ethics and law – this series will critically examine how science and technology are shaping the emerging international political economy.

Published titles in the series:

The Complexity Challenge by Robert W. Rycroft and Don E. Kash

Developing Innovation Systems by Mario Cimoli

Evolutionary Economics and the New International Political Economy edited by John de la Mothe and Gilles Paquet

Global Change and Intellectual Property Agencies by G. Bruce Doern

The Governance of Innovation in Europe by Philip Cooke, Patries Boekholt and Franz Tödtling

Regional Innovation, Knowledge and Global Change edited by Zolton ACS

Services and the Knowledge Based Economy edited by Ian Miles and Mark Boden

Systems of Innovation edited by Charles Edquist

Universities and the Global Knowledge Economy edited by Henry Etzkowitz and Loet Leydesdorff

Proposals for books can be sent directly to the series editor:
    John de la Mothe
    Program of Research on Innovation, Management and Economy (PRIME)
    Faculty of Administration
    University of Ottawa
    275 Nicholas Street
    Ottawa, Canada K1N 6N5
    E-mail: delamothe@admin.uottawa.ca

Science, Technology and Governance

Edited by
John de la Mothe

Routledge
Taylor & Francis Group
New York   London

Science, Technology and the International Political Economy Series
Series Editor: John de la Mothe

First published 2001 by Continuum.

This edition published 2013 by Routledge

711 Third Avenue, New York, NY 10017
2 Park Square, Milton Park, Abingdon, Oxon OX14 4RN

Routledge is an imprint of the Taylor & Francis Group, an informa business

British Library Cataloguing-in-Publication Data
A catalogue record for this book is available from the British Library

ISBN 0–8264–5027–X (hardback)
     0–8264–5026–1 (paperback)

Library of Congress Cataloging-in-Publication Data
Science, technology, and governance / edited by John de la Mothe.
     p. cm. — (Science, technology, and the international political
     economy series)
  Includes bibliographical references and index.
  ISBN 0–8264–5027–X — ISBN 0–8264–5026–1 (pbk.)
  1. Science and state. I. De la Mothe, John. II. Series.

  Q125 .S43473 2001
  338.9'26—dc21
                                                              00–052343

Typeset by YHT Ltd, London

Contents

Contributors

Lewis Branscomb is Atena Professor of Public Policy and Corporate Management (Emeritus) in the John F. Kennedy School of Government at Harvard University.

Mario Cimoli is a Director at ECLAC, United Nations, Division of Production, Productivity and Management in Santiago, Chile.

Susan E. Cozzens is Professor and Chair of the School of Public Policy at the Georgia Institute of Technology.

John de la Mothe is Professor of Science and Government in the Program of Research in Innovation, Management and the Economy (PRIME), Faculty of Administration, at the University of Ottawa, and Visiting Professor at Yale University.

Francis Fukuyama is the Omer L. and Nancy Hirst Professor of Public Policy at George Mason University, Fairfax, Virginia.

Michael Gibbons is Secretary General at the Association of Commonwealth Universities in London.

Philip Gummett is Director of Higher Education Funding Council in Cardiff.

David M. Hart is Associate Professor at the John F. Kennedy School of Government at Harvard University.

Sanford A. Lakoff is (Emeritus) Research Professor of Political Science at the University of California at San Diego.

Jane Marceau is Pro-Vice-Chancellor (Research) at the University of Western Sydney.

Judith Reppy is the Associate Director of Peace Studies at Cornell University.

Jean-Jacques Salomon is Honorary Professor of Technology and Society at the CNAM, Founder of the International Council for Scientific Policy Studies, Paris.

Eugene B. Skolnikoff is Professor of Political Science (Emeritus) at M.I.T.

Luc Soete is Professor of International Trade and Economics at MERIT, University of Limburg.

Caroline Wagner is a Director in the Science and Technology Policy Institute at the RAND Corporation, Washington, DC.

Preface

The interface of science, technology and public policy presents a rich, demanding and important area for study.

It is rich and important, in part, because so much of contemporary life and potential is mobilized or challenged by science, technological innovation or research. The ebbs and flows of the NASDAQ, the rise of 'smart communities', and the issues of xeno-transplantation, cloning and biodiversity all highlight this richness of science, technology and public policy, as do issues of employment, development and environment.

'Science policy' – to use the shorthand – is a demanding area of study because it is by definition multidisciplinary. Professionally, it therefore stands outside, or only uncomfortably inside, the confines of discipline-based colleges, universities and academic careers. It is demanding also because it requires an understanding of many methods, approaches and perspectives – an appreciation of lab life as well as of economics, sociology, history, politics and the machinery of government. It is also, in its deepest sense, a field of cultural inquiry, revealing shifts over time in human values and ethics.

But such inter-linkages – which are essential and central to all human activity – are not the norm in many academic pursuits. Instead, they need to be fought for. Students of economics and management rarely are asked to go into a lab to observe researchers working on nano-technologies in the development of biomechanical devices, or surface chemistry or fuel cell technologies. Students of the natural sciences are still, regrettably, taught the myths of the 'scientific method' – which has been criticized pedagogically at least since the 1920s – and rarely are they encouraged to study not only how they do science but what science does and means, and why. And students of politics too often view science and technology as both intangible and as just being ephemeral to their interests, unlike political questions of homelessness, airport locations, or industrial development for which controversy and photo ops can more clearly fit political agendas and analysis.

And finally, the study of science policy is important because science, technology and democracy represent human activities that are perhaps the most powerful forces humankind has ever invented.

So in a sense this book is about the practice and thinking behind science, technology, innovation and their interfaces with government. It is about the governance of knowledge, the governance of ideas. In a sense these themes are key because the nature of government and governance is shifting. In reading this volume, we hope that the myths of science policy will become more transparent, that the practice of science policy will be seen as being atheoretic (or at least as pluri-theoretic), and that the issues that are emerging from the knowledge–governance interface will be viewed as challenging and deserving of our attention and action.

This volume has taken longer than expected to develop. In its preparation I would like to take the editor's prerogative to thank a number of institutions and individuals. First, thanks go to the Social Sciences and Humanities Research Council of Canada and its Innovation Systems Research Network, of which PRIME at the University of Ottawa is a part. Second, thanks go to my students in the Faculty of Engineering and Applied Science at Yale University who generously beta-tested an early draft of these chapters. In particular I would like to mention the help of David Buchwald, Michael Burstein, Bella Desai, Bill Grady and Alex Selkirk. Third, I would like to thank Anastasia Vakos of Queen's University, Kingston, Ontario (Finance and Economics) and Tyler Chamberlin of the University of Ottawa (Economics and Public Policy) for making the physical text a reality. Fourth, I would like to note that the chapters by Michael Gibbons, Luc Soete and Jean-Jacques Salomon were initially offered as PRIME Public Lectures at the University of Ottawa, while that by Lewis Branscomb was offered as a joint PRIME-PAGSE (Partnership Group in Science and Engineering) lecture at the National Press Club in Ottawa. My thanks thus go to Howard Alper of PAGSE for this collaboration. Of course, it would be remiss of me if I did not thank Paul Dufour and Fred Gault for always lending me their critical eyes and kind support, as well as Nora O'Neil of Harvard's Kennedy School of Government for bibliometric reassurances, now and over the years. Finally I would like to thank the contributors to this volume – friends and colleagues all – both for their insights and their patience.

John de la Mothe
New Haven,
Connecticut

PART I

Introduction

CHAPTER ONE

Knowledge, Politics and Governance

John de la Mothe

Introduction

The governance of knowledge, ideas and creativity is one of today's powerful leitmotifs. But by examining the socio-economic dynamics of science, technology and innovation through the lens of 'governance', we are not speaking narrowly of 'policy', although that is centrally subsumed. Nor are we arguing directly either for more or less government intervention, although this is a common theme when the Belt Way, Whitehall, the Hill or other 'Corridors of Power' stumble through political and economic debates between the traditional 'left and right', between intervention and *laissez-faire* – as *passé*, yet lingering, as they are. Nor are we heralding any linear models of innovation which relegate science policy to either the funding of scientific research and technological development, or to the performance of research in support of government policy. Instead, we see the complex of activities that is casually referred to as 'science policy' as being deeply integrated into the fabric of our modern societies and economies. Together these activities are the signature activities of economic growth, productivity, wealth creation and distribution, of environmental sustainability, and of improved and expanded health, welfare and security.

By talking about governance, we are not talking about top-down decision-making since we have moved beyond traditional notions of hierarchy and authority. Instead, we are talking about joint steering. 'Governance' is about the handling of complexity and the management of dynamic flows. It is fundamentally about interdependence, linkages, networks, partnerships, co-evolution, and mutual adjustment. Thus we have transcended such high level and highly stylized dichotomies as Science versus Politics (in which – as Sir Peter Medawar (1967) put it – the objective art of the soluble and the subjective art of the possible could not mix). In the style of Karl Polanyi and Kenneth Boulding, as well as such cultural anthropologists as Mary Douglas, our focus in this volume is with the continuous interaction and negotiation between institutions and networks; between balancing and maximizing economic opportunity, political stability and social cohesion. In this sense, thinking about science and technology policy in terms of governance draws us inevitably into the realm of evolutionary economics – as enunciated by Christopher Freeman, Richard Nelson, Sidney Winter, J. Stanley Metcalfe, Giovanni Dosi and others – as well as into the periphery of systems thinking à la Robert Jarvis and Stephen Kline and directly into the headlights of Joseph Schumpeter and F. A. von Hayek. We are also forced to deal with the tacit dimension of knowledge, not only its codified aspects, as riveted by Michael Polanyi, and to reconsider socio-economic activity in terms of intangibility, information and services, as Ian Miles and Johnathan Gershuny have done. Thus the governance of science, technology and innovation is not about centralized mindsets and bureaucracies but is instead about finding the optimal level of decision-making and dispersion, power and legitimacy, participation and action as it pertains to the new production, use and diffusion of knowledge – our most important currency.

The unfinished project

Knowledge production, identification, application and distribution is a never-ending process of 'standing on the shoulders of giants', as C. P. Snow once put it referring to the days when the physicists 'changed the world' (1981). These processes, and the activities that encourage them to happen to the benefit of all, continually remind us that knowledge is, indeed, power.[1]

This slogan has been central to our culture's architecture ever since Francis Bacon (1561–1626) bequeathed it to us in the form 'Scientia est Potentia'. Acting as a focal point for the Enlightenment with its commitment to Reason and Progress, which Jürgen Habermas and other Critical Theorists have reviewed as our 'unfinished project' (Marcuse 1966) and drawing on the questioning spirit of Thomas More's Utopia (1618) – ironically, a term which means 'nowhere', Bacon's motto was to be adopted by the Royal Society of London, passed on through the 1933 Chicago World Fair, and landed as the *de facto* motto of our knowledge-based economy. This deep commitment to liberalism (both economic in the European sense and neo-conservative in the American sense) and Progress led us to a preoccupation with instrumental reason, an attempt to perfect our tools. It also led to the arrogance of positivism, and, some would argue, to the imbalance between instrumentality and wisdom. none the less, Bacon's dream *did* give shape to the rise of the gentleman amateur, the natural philosopher, the organized industrial lab, 'invisible colleges', peer review, mass literacy and the rise of the scientific journal, scientific societies, traveling itinerant lecturers, evening classes for mechanics, the development of academic disciplines, and in their wake a wide range of social, environmental, and economic gifts through scientific advance, technological development, and the application of innovations.

However, the advance of Reason was only one part of the equation underlying what we now call 'science policy'. The other side of modern scientific development has continually been the rise of democracy. Building on Bacon, particularly his *New Atlantis* (1626) which was a brilliant exposition 'on the rise of science in an age of superstition', 'the Marquis de Condorcet synthesized the Baconian conception of the utility of science with the Rousseauistic and Lockean vision of a society of freedom and equality. Condorcet believed that the progress of freedom and the progress of science were two sides of the same coin: "Nature has joined together liberty, virtue and respect for the natural rights of man". In order for there to be scientific progress there would have to be freedom – freedom from censorship, dogma, superstition and fanatacism – and in order for there to be freedom on the universal scale the sciences would have to progress to such an extent that the gap between the "haves" and "have nots" would be overcome' (Lakoff, 1966: 16). Still today we face both new and long-standing gaps in the distribution of welfare, but the quest to narrow these gaps continues on many fronts.

Herein we find the modern dilemma buried in a spectrum of almost ideological choices. Bacon's dream – a promogenitor of Vanevaar Bush's 1946 blueprint which led to the deliberate establishment of funding structures and institutions (such as the National Science Foundation which is seen by many as being the embodiment of Bacon's 'Solomon's House') – leads *both* to Michael Polanyi's 'Republic of Science' which was deeply anti-collectivist (and therefore vehemently opposed to any form of centralized – government – planning for science) *and* to J. D. Bernal's 'social function of science' through which the utility of science could be planned by government and its benefits distributed. This essential tension is often seen in the transformative practice of science and technology policy itself in which governments increasingly phrase their options today in terms of 'innovation systems' (i.e. in terms of an integrated supra-policy format) and in a growing recognition of the realities of now working increasingly in a Mode II (transdisciplinary) context, as opposed to Mode I (Gibbons *et al.*). Questions of how much public funds should be invested in research and development (R&D) are quickly followed by questions of what fields should be invested in, and what institutional mechanisms – such as block grants, strategic grants, leveraged or matched funding with industry, individual or consortia grants, and so on – need to be designed or used.[2] As an evaluative tool, governments often use such macro-indicators as the Gross Expenditure on R&D (GERD) as a percentage of Gross Domestic Product (GDP) as a proxy for the relative competitiveness of the nation.[3] In other words, having accepted the view that investments in science and technology pay social and economic dividends, then the practice of government becomes occupied with support, strategy, and priority setting.

The politics of knowledge

Culture

But we must remember as a backdrop that such a belief in science is not universally accepted. Part of this disquiet has been genteelly defined in cultural terms.[4] C. P. Snow's *The Two Cultures and the Scientific Revolution* is perhaps best remembered in this regard. But other works – for example by Jacob Bronowski, Aldous Huxley and J. Robert Oppenheimer – have long echoed concern about a cultural divide, not so much between the arts and the sciences but more between scientists and the rest of the population. As Sir Herman Bondi noted in his preface to Bronowski's *The Common Sense of Science* (1979) 'in 1950 the general public thought of science in this way: "incomprehensible" ... [In] 1970 this had changed to the following view of science: "incomprehensible".' Today this is probably overstated – although some promoters of the public awareness of science would disagree – with so many people making their livings in high technology industries and using technological tools. But the disquiet remains, and we must recall that the vast majority of policy decision-makers do not have scientific, engineering, medical or research backgrounds, thus making the continuing case for arguing on behalf of continued and expanded funding important.

Risk, accountability and inclusion

Beyond cultural debates, a second motif can be found in the areas of risk, accountability and inclusion. Recall the powerful impact of Rachel Carson's *Silent Spring* and of Ralph Nader's *Un-Safe At Any Speed*. During the 1960s and 1970s, proper public concern was focussed and expressed around environmental degradation, toxic spills, aeroport citing decisions, and nuclear safety and responsibility. These concerns have been articulately documents over time by such scholars as Dorothy Nelkin and Sheila Jasanoff.

But underlying this class of risk and accountability concerns, as suggested by John Kenneth Galbraith in *The New Industrial State* (1967), is in fact a hydra of concerns – often expressed in terms of a general suspicion of multinational enterprises and of experts. In Europe the foreign MNE issue was captured by Charles de Gaulle's complaint about 'American technological imperialism' while in America, bumper-stickers declaring 'do not bend, fold or mutilate' dove-tailed with an anti-Vietnam, anti-corporatist, pro-individualistic 'Woodstock' sentiment. But again here, this nest of concerns has shifted with both the tremendous globalization of industry, the vertical and horizontal integration of multinationals, the growing public awareness of risk, and the emergence of non-governmental organizations dedicated to the protection of myriad environmental and health issues. In a governance framework, we can see the re-balancing of institutional life in what Susan Strange called 'the retreat of the State'. Here again we can see the changing relationships between Science and Politics.

In terms of accountability a concern is that scientists and engineers – who as a class were often referred to in the 1960s as 'the new priesthood' and 'the new Brahmin' – are isolated or remote (Klaw, 1968). On this point, and using a specific example often used by Bertrand Russell, 'can a mathematician who contributes to the design of an explosive device be held morally responsible for injuries caused by the use of the device, even though he or she was unaware of the purpose of their calculations?' This moral dilemma is a theme that emerges squarely in C. P. Snow's 1954 novel *The New Men* and is a question which haunted Albert Einstein whose own mathematics is seen by some as having led directly to the atomic age. For Russell, the answer was fairly straightforward: there is always a moral choice in what to do next. As fascinating as this philosophical and practical debate can be, this sort of perceived dissociation between scientist and the public good fuels antiquated and erroneous views of scientific knowledge as being 'objective' and 'true'. But in today's practice researchers are not solo agents, and they do not work at odds with public needs in a Frankenstein way, but are in fact deeply accountable institutionally – to their deans, departmental chairs, vice-presidents of research, other corporate managers, grant selection committees, funding agencies as well as to their professional associations.

But concern with risk has provided a valuable platform for the involvement of people in technical decisions. Today for example passionate debates that involve genetically modified foods, food additives, cancer maps, xenotranplantation, and 'smart viruses' have brought experts, policy-makers, industrialists and citizens together in the search for responsible and progressive research directions and controls. So again, along with the cultural arguments noted above, the relationship between Science and Politics, or between science and government, has evolved. No longer is an 'us–them' adversarial framework of any utility, and it certainly does not improve science policy processes. Indeed, what we have seen since Bacon is a development of science and technology from a phase of *understanding Nature* (complete with all of its taxonomic work as in the Periodic Table of elements and the Messier Catalogue of stars) to *controlling and using Nature* (as in the Scientific, Industrial and Information Revolutions), to today's perception that we are about to *change Nature* through cloning, gene therapy, and other developments in the biological and bio-informatic realms. A new arena of risk and opportunity is dawning and, through a governance framework, all sectors and interested citizens can be engaged to promote continued technological developments and responsible management and monitoring. Together, this class of tides – risk, accountability and inclusion – represents a widening recognition of the need for a new social contract between science and society. The emerging class of common and trans-boundary (i.e. international) problem areas will require our pursuit of Mode II solutions.

On the point of inclusion, as the June 1999 UNESCO world conference on science in Budapest revealed, there are large groups which, in search of a new social contract and wishing to share in the benefits of science and technology, are seeking a broadening of representation in research, development and in the conception of reliable knowledge. These groups, representing developing nations, women's groups, aboriginal and other traditional forms of knowing (Polanyi, 1957), made an articulate case that was dutifully reported in the pages of *Nature*.

Social Relativism

But this search for legitimate inclusion and a broadening of Western notions of reliable knowledge does not fall into the final issue area in the politics of knowledge through which we can see a growing social relativism of scientific knowledge to the extent that science – in some circles, but sanctified by legislation – is just another belief system. This is wrong – but this reaction to modern knowledge is in part the result of the academic Left, particularly in France and the United States, deconstructing knowledge and applying postmodern 'norms' to meaning and 'speech-acts' (including research results). In a celebrated debate, initiate perhaps by a book by Paul Gross and Norman Levitt entitled *Higher Superstition: The Academic Left and its Quarrels With Science* (1994) and given voice at a New York Academy of Sciences meeting, published in 1996 under the title *The Flight From Science and Reason*, the argument was put thus: 'Evidence of a flight from reason is as old as human record-keeping . . . . But *rejection* of reason is now a pattern to be found in most branches of scholarship and in all the learned professions' (emphasis added). That such a trend exists is worrisome and is more mischievous than the cultural concerns noted above of C. P. Snow. But what is more dangerous is the idea that research results are mere opinion, and that 'experts' are not longer 'certain'. Scientific uncertainty has long been part of public debates. In the case of Climate Change for example, scientists legitimately still cannot tell policy-makers what to do because – in such a complex and multidisciplinary area of enquiry – we cannot say 'for sure' whether shifts in weather patterns are anthropogenic or not. Another example might be the debate around the Canada–US Bilateral Agreement on Acid

Rain in the 1980s in which representatives from New Hampshire and Ontario stymied the negotiations because experts could not say 'for sure' what increase in acid rain in the north-east US would be the result of increased automobile use in Toronto. The agreement finally became focussed on the possibility of lost jobs rather than on environmental degradation. Thus to science policy analysts, dealing and trading with scientific and political uncertainty is par for the course. But to a less engaged public, the exchange eroded classical beliefs in the veracity of science. More seriously, in watching genetic experts pitted against one another by lawyers on television, as in the O. J. Simpson case, panders not only to the lowest form of entertainment but fuels the views of those who would like to insist that science can only be taught in equal time (if at all) alongside religious beliefs in school as has been done in a number of US state school boards. This is not just silly. It masks some dangerous elements.

Governance in a Mode II society

Clearly, as the above has sketched, science and technology have both contributed greatly to our common well-being. But their authority is not without contention – they are not without 'a politics'. Yet as this book makes clear, traditional frameworks for the politics of knowledge are being radically reconfigured. Science and technology policy are increasingly being integrated into all aspects of policy – from trade and investment, to foreign affairs, to health and safety, to economic and environmental concerns. Additionally, science and technology are accelerating in presence and impact – on job and industrial structures, on globalization and the interdependence of nations, on consumer behaviors and market choices. Through telecommunication, information and nano-technologies, knowledge-intensive products and services are becoming embedded in everyday life. We have entered an age quite unlike the age of mechanical technologies. Of that era, the cultural and literary scholar Lionel Trilling could write that the Wright Brothers' machine which they flew at Kitty Hawk defied comprehension but it 'had nothing to hide – you could see every moving part'. Today this is not true. Technology has become invisible, just as Fumio Kodama has been able to say that international corporate competition in a global economy now features 'invisible enemies' – it can come from any sector, or any size firm or nation (Kodama, 1991).

Hence – beyond firmly understanding the changing nature and structure of science, research and development (which is key) – the challenge of governing science and technology really has become a task of focally (Polanyi, 1957) dealing with complexity, immediacy and speed (Rycroft and Kash, 1999). We need to develop learning institutions which are flexible, adaptive, responsive, and close to their 'clients' (be they suppliers, final customers, or citizens). Transparency and horizontal coordination demands a dispersion of decision-making power as well as a true commitment to consultation and transparency and not (as Jean-Jacques Salomon puts it so elegantly at the close of this book) just with offering the chance to 'participate in participation'. This requires a commitment to the development of trust and 'social glue'.

At the outset of this broad prolegomena, I noted the deep historical and cultural connections between science, technology and democracy. These connections are alive and well, although – let us not kid ourselves – they are continually challenged and are sometimes packaged in terms of trade liberalization or the open society. This book deals with these themes. It should be noted however that this is not the result of a joint project but is instead a conversation between a few 'fellow travelers'. The intent of this volume is therefore not to forward a thorough-going treatment of the relationship between science and government, as path-breakers such as Don K. Price (*Government and Science*, 1954), J. Stefan Dupré and Sanford A. Lakoff (*Science and the Nation*, 1963), and Harvey Brooks (*The Government of Science*, 1968) have done thirty (and more) years ago. Instead, it is intended to provide a fairly comprehensive tour of the issues – extant, embedded and emergent – that science policy-makers, students and analysts will have to deal with now and in the near future, to encourage a broadening of attention and accessibility to S&T discussions, and to highlight certain tools that are available.

Notes

1. I am reminded of the bumper sticker I saw in Cambridge, Massachusetts, near the campus of MIT which read 'if you think education is expensive, try ignorance'.
2. In Canada for example, recent institutional innovations include the creation of the Canadian Foundation for Innovation (CFI) alongside the long-standing Natural Sciences and Engineering Research Council as well as the replacement of the Medical Research Council with the Canadian Institutes for Health Research. Matching grants, introduced in 1986, began the move towards leveraging private sector funds into R&D. Later, Networks of Centers of Excellence (NCE) were introduced as a way of linking researchers across the country.
3. Other typical indicators deal with R&D in the higher education sector and in business enterprises, productivity, patents, and the technological balance of payments. The central texts guiding these statistically based indicators are the OECD's *Frascati* family of manuals. The National Science Foundation collects and publishes an extensive array of S&T indicators, as does the European Commission.
4. The contributions of F. R. Leavis were not so genteel and almost resulted in a lawsuit. See John de la Mothe, *C. P. Snow and the Struggle of Modernity*, Austin University of Texas Press, 1992.

PART II

Shifting Dynamics in Science and Government Relations

CHAPTER TWO

Politics and the Communities of Science

Sanford A. Lakoff

Introduction

In 1663, the chemist Robert Hooke advised the fellows of the Royal Society of London to avoid 'meddling with Divinity, Metaphysics, Moralls, Politicks, Grammar, Rhetoric or Logick'. This notion, that scientific speculation and experimentation could be kept separate from other concerns, would survive – but only until scientists became too successful to be left to their innocent pursuits. Except for religious concerns with cosmology and the theory of evolution, those who investigated such matters as the motions of the planets and the taxonomy of living things were generally not hindered from doing their work or communicating with each other, even in times of war (DeBeer, 1960). But there were hints of what was to come. At the same time as Hooke made his plea, Francis Bacon was trying to win support for inductive science by contending that the 'Great Instauration' was no threat either to faith or to political order but was only a new instrument – *a novum organon* – for the relief of man's estate (Lakof, 1966). The very utility of science that Bacon foresaw was to spell the end of its autonomy.

Now that the advance of knowledge is not only the source of countless practical applications for 'the relief of man's estate' but is the decisive index of modernization and military power, the old separation of science from social concerns – and with it the relative independence of the 'republic of science' (Polanyi, 1962) – has become impossible to maintain. The fruits of scientific research are too important to national security, and to economic growth, welfare, and public health, for scientists to escape all social direction and regulation, and the by-products of industrial civilization have so much effect on the safety and balance of the environment that regulation of technology and remediation of its harmful effects must be essential instruments of government. At the same time, scientists have become heavily dependent upon governments, directly and indirectly, for their livelihood and for their very ability to do research. Especially in the case of 'big science' projects, with uncertain and remote pay-offs, public subvention is a *sine qua non*. On all these scores, scientists are inevitably drawn into the political process as technical advisers, policy adversaries, and supplicants for support. While most continue to prefer their laboratories and classrooms to parliamentary lobbies and administrative committees, many have come to play a prominent role in political affairs, both domestic and international – some have done so with a sense that the bearing of their work imposes a special civic responsibility (Hawkins *et al.*, 1986; Lakoff, 1979).

This great change in the relationship of scientists to society did not come about initially because it seemed mutually beneficial. Rather it emerged in reaction to an enforced politicization of science. Totalitarianism, both as a source of ideology and as a system of social control, was the serpent that first drew scientists out of their own state of nature. Scientists in the Soviet Union were caught up in the purge of the old intelligentsia ordered by Stalin. They suffered not only philosophically from his insistence that dialectical materialism be accepted as the foundation of all knowledge, natural as well as social, but more seriously from a host of constraints that made their personal and professional lives difficult and sometimes impossible to maintain. In place of the old cosmopolitanism and openness of the scientific enterprise, they had to endure enforced isolation from 'Western' or 'bourgeois' science. At Stalin's direct order, the physicist Pyotr Kapitza was forbidden from returning to the Cavendish Laboratory in Cambridge, England in 1934 (Badash, 1985). (Rather than rebel or go into 'internal exile', however, Kapitza accepted political discipline and worked within the system to direct a research program which eventually, with no small help from espionage, produced an atomic bomb only four years after its development in the United States (Holloway, 1994).) Stalin also gave official approval to the work of the pseudo-scientific 'agronomist' Trofim D. Lysenko. At Lysenko's instigation, the 'orthodox' geneticists were persecuted and their leader, S. I. Vavilov, who was also the head of the Soviet Academy of Sciences, was arrested and forced to concoct a bogus confession admitting that he had engaged in sabotage (Graham, 1967; Joravsky, 1961b; Josephson, 1991; Krementsov, 1997). In Nazi Germany, racist ideology led to the expulsion of 'non-Aryan' scientists. By a conservative estimate, a quarter of German physicists were forced to leave the country, including Albert Einstein and nineteen other Nobel Laureates. Those who remained were expected to serve the new regime and its goals loyally and did so, a few out of ideological sympathy, most out of a combination of patriotism and the hope of protecting the country's scientific institutions (Beyerchen, 1977: 47, 63). All who remained were willing to overlook the evil character of the regime and some physicians conducted grossly inhumane medical experiments on concentration camp inmates (Cole, 1983; Haberer, 1969). In reaction, scientists in the democracies began to impose self-censorship on their publications in sensitive areas of research, and readily turned their talents towards war preparedness.

The two world wars of the twentieth century, especially the second, served as the strongest catalysts for change in the relationship of scientists to the state. World War I brought some scientists into war-related research and divided the international scientific community by nationality. Once the war was over the old bonds were restored and a golden age of cooperation, especially for physics, made a brief but shining appearance (Kevles, 1977; Schroeder-Gudehus, 1973). World War II mobilized scientists in war-related projects and produced rifts that were to prove more lasting than those of World War I.

In 1945, at the end of World War II, the one major power whose geographical isolation had kept it largely unscathed was the United States. It was here that the changing relationship first took shape. After the war was brought to a sharp and decisive end by the use of the atomic bomb, the 'absolute weapon', political leaders needed little convincing to agree that superiority in science and technology would henceforth be essential to national security and international power (Brodie, 1946). Some analysts claimed that the bomb had actually been used to prevent the entry of the Soviet Union into the war against Japan, and was therefore in effect a prelude to the Cold War (Alperovitz, 1965). At the urging of such wartime scientific administrators as Vannevar Bush and James Bryant Conant, the federal government became the principal patron of science (Bush, 1945). Universities were allowed considerable autonomy in conducting basic or curiosity-driven research, with support provided via the peer-reviewed project grant system. For development, the federal government relied on privately owned corporations through somewhat more restrictive but financially more generous negotiated 'R&D' contracts. The lion's share of the funding came from the Department of Defense and the other mission-oriented agencies. The National Science Foundation was established in 1950 in order to support basic research in the physical sciences and monitor the nation's scientific activity. The existing National Institutes of Health were greatly enlarged and given more financial support to support the life sciences and search for cures to major diseases. The National Aeronautics and Space Agency (NASA) was set up to manage space exploration, the Atomic Energy Commission (AEC) to control all work on nuclear power. Through these federal agencies and programs, a remarkably productive effort in science and technology sprang up around the research universities and in the high-technology industry that catered to the military. By the mid-1960s, the country was devoting some three percent of its Gross Domestic Product (GDP) to the support of R&D, a figure that became a benchmark for emulation by other countries even though it was considerably inflated because of its military character (Brooks, 1968; Dupré and Lakoff, 1962). An observer with a keen sense of American history described the country's pragmatic and pluralistic approach to science as 'federalism by contract' (Price, 1954). One critic (Nieburg, 1966) saw it as the rise of 'the contract state', another as 'Pentago capitalism' (Melman, 1970).

In response to the 'Sputnik crisis' of 1957, the United States established an office of Special Assistant to the President for Science Technology, subsequently downgraded but revived as the White House Office of Science and Technology Policy (OSTP). The office was designed to provide the President with scientific advice, with the help of panels appointed by the President's Science Advisory Committee (PSAC) and to work in tandem with the Bureau of the Budget (subsequently renamed the Office of Management and Budget) in setting priorities and reviewing the intramural work and extramural grants and contracts of the federal agencies with responsibilities in science and technology. The National Academy of Sciences-National Research Council, an elite organization chartered by Congress, was called upon for extramural advice (Boffey, 1975). A Federal Council for Science and Technology was created to serve as a kind of sub-cabinet for science and technology which would allow for greater coordination among the various departments with heavy investments in R&D. The Food and Drug Administration (FDA) was given new powers and an expanded capacity to regulate the introduction of new drugs. One of its regulators, Frances Kelsey, blocked the introduction of the drug Thalidomide in time to prevent the birth of deformed children. In both houses of Congress, new committees were struck to monitor government science activities, including a rare joint committee to oversee atomic energy. In 1970, the

Environmental Protection Agency (EPA) was added to the structure in response to growing concern for the safety of the natural habitat, and a cascade of environmental legislation soon followed, designed to clean the air and water and regulate the use of insecticides and even noise. In 1973 Congress passed the landmark Endangered Species Act. In 1980 a 'Super Fund' was created to direct and assist the clean up of toxic waste sites. As these developments took place within government, organizations of scientists outside government began to play a more prominent role as advocates of support, purveyors of advice, and in struggles over many policy goals with scientific-technological content, including nuclear testing and arms control and increasingly, such issues as family planning, the effects of pesticides on the environment, the safety of food additives, and the effects of smoking and health.

Other countries followed suit, though, with the significant exception of the Soviet Union, civil rather than military science was the predominant concern (Berry, 1988). In 1958, for example, Germany established a Wissenschaftsrat or Science Council and the cabinet post of Minister for Science and Technology, to share responsibilities for science policy with the Minister of Health and Welfare (Meyer-Krahmer, 1990). Japan also sought to assure cabinet-level coordination of science policy by creating a Council for Science and Technology of which the Prime Minister is chairman. In addition, each cabinet department has its own scientific advisory council. Major responsibility for organizing and stimulating industrial innovation was assigned to the Ministry of Industry and Innovation (MITI) (Anderson, 1984). Britain already had in place Research Councils in charge of distributing formula-based funds for basic research to the universities. The post of Chief Scientific Adviser to the Cabinet was added, along with an Advisory Council on Science and Technology (ACOST) (Cunningham, 1999; Nicholson *et al.*, 1991). France, in an effort to blunt American expansion and restore its pre-war role as a major European power, created its own Atomic Energy Commission, the Commissariat a L'Energie Atomique and initiated its own atomic weapons program. In the 1950s, a secretariat, responsible to the Prime Minister was appointed. This first step led to the establishment of a government agency, the Délégation Générale à la Recherche Scientifique (DGRST), which set out to strengthen the links between the Centre Nationale de Recherche Scientifique (CNRS) and the universities (Gilpin, 1968; Rouban, 1988: 110–22). The function continued to be performed, but with lower governmental status, until, in 1997, a newly elected prime minister created a Ministry of National Education, Research, and Technology and appointed a geochemist to serve as its head. Canada too sought to strengthen the links between the federal government and the research communities. On the advice of a Senate commission, the post of Minister of State for Science Policy without portfolio was added to the Cabinet, supplemented by a federal Science Council (since dissolved as an economy move) to perform research on national science policy (de la Mothe, 1992; de la Mothe and Dufour, 1993; Doern, 1972; Lithwick, 1969.) Virtually every other country in the world has since adopted similar measures, though they have not all survived in original form (Ferné, 1989; Forje, 1989; Golden, 1991; Kellerman, 1988; Ronayne and Boag, 1989; Solingen, 1994). The most recent to do so have been the countries of Eastern Europe, following the collapse of communism (Darvaes, 1988; Popper, 1991).

In the developing countries, the end of colonialism led to efforts to build homegrown institutions of science and technology, in keeping with the 'import substitution' model adopted by most as the strategy for economic growth. These efforts have faced special difficulties due to the scarcity of resources, poor educational infrastructure, the weakness of central authority, the 'brain drain' of talented young people to the advanced countries, and opposition or at least suspicion from entrenched traditional attitudes, including the belief in folk medicine. In most of these countries, efforts to nurture science also had to overcome deep-set biases such as those of caste and gender in order to make the fullest use of human resources. In Mainland China, scientists were among the traditional elites subjected to criticism and persecution during the years of the 'Cultural Revolution'. Despite these difficulties, great progress has been made. In general, the strong desire for modernization on the part of ruling elites, whether civil or military, has helped legitimize science and provides significant support for technology, if not always for basic research. Major national scientific figures, like Abdus Salam in Pakistan and Homi Bhabha in India, have given special impetus to the programs in their respective countries. In the 'Asian tigers' (notably Taiwan, Singapore, South Korea, and Hong Kong) and the most successful Latin American developing countries, keen concern for high-technology-driven export industries has inspired investment in research facilities, greatly helped by technology transfer from multinational corporations. International scientific organizations such as the International Council of Scientific Unions (ICSU) have set up special working groups to help the developing countries (Jones, 1971; Latin American Newsletters, 1983; Rahman, 1990; Solingen, 1994).

Despite these virtually universal institutional changes, it would be wrong to suppose that the entry of scientists into the political arena has altogether changed the character of the political process, either in the advanced or developing countries. While scientists and technologists are relied upon for work vital to the state, and are sometimes consulted on policy matters, they do not exercise a decisive influence on most public questions. Few scientists and engineers have held or aspired to political office. Political questions continue to be decided by constitutionally authorized officials, or, in authoritarian settings, by those with de facto control. In democracies, elections are still fought by political parties in campaigns in which scientists have only a marginal role in their professional capacities. Before and even during World War II, 'scientists did not yet perceive themselves as powerful figures of domestic politics. They regarded themselves and were regarded by political leaders as more spectators than participants in political affairs' (Beyerchen, 1985: 69). Since then, scientists have certainly become more prominent in public affairs, but like the other professional and technical groups with which they overlap, they are for the most part only bit-players in the political drama, even though their work is increasingly crucial to state policy.

In some important respects, however, scientists have become more than just another constituency to be placated or ignored in favour of numerically larger and better organized groups. The main reason they cannot be ignored or treated cavalierly is that many of the key substantive issues that engage public attention – nuclear weapons and the safety of nuclear power, the need for alternative sources of energy, strategies for dealing with epidemics like AIDS, the fear of global warming, as well as such bioethical issues as contraception, abortion, euthanasia, and the effects of cigarette smoking – arise because of the progress of science and require the advice and implementation of scientists. As a new 'strategic elite' (Keller, 1963), they are in a position to influence decision-making in many matters where their expertise is called upon. Even the language and to some extent the method of science have influenced political discourse and the manner of reaching and rationalizing decisions, leading to the more and more widespread use of statistics, models, projections, simulation and 'systems analysis'. As issues with scientific content become ever more salient, and to the extent that the general public becomes better educated, scientists can be expected to play a still greater role. A countervailing tendency has also developed, however (Ravetz, 1971). Fear of catastrophe has led to anti-technological and sometimes anti-scientific protest movements directed against nuclear power, genetic manipulation, animal experimentation, and, on religious grounds, continuing opposition to the teaching of evolution. Scientists and technologists therefore sometimes find themselves in the role of an embattled professional elite confronted by a populist backlash.

Some prominent or simply committed scientists have become involved in politics as individual spokesmen, crusaders, or advocates. Many more have done so as members of professional associations or in collaboration with lay groups concerned with policy issues. Conversely, governments have taken a growing interest in scientific activities, in matters ranging from the 'cardinal questions' of life and death (Clark, 1965; Snow, 1981) to such more mundane matters – but vital to the health of science – as what percentage of national budgets should be devoted to supporting research and development and higher education (Brooks, 1968; Greenberg, 1967; Smith, 1980). As a result, science and government are linked in myriad ways, notably in the following four main areas of concern.

Warfare

As applications of science have revolutionized the arts of war, those responsible for the development and assessment of new military technologies have acquired at least an instrumental role, and often a consultative one, in policy-making for national security and arms control. In open societies, they have been prominent in debates over strategy, the development and adoption of particular weapons systems, and prospective arms control measures. They have served as diplomats and as advisers to diplomatic delegations in negotiations over arms control (Jacobson and Stein, 1966; York, 1987). Some do so out of a sense of moral responsibility for the uses of their work (Lakoff, 1966; Reid, 1969), others because their expertise is considered necessary in policy discussions. In democracies, the great majority of scientists accept the principle that policy matters should be decided by citizens and their elected representatives. In matters of policy, they see their role as that of providing politicians and voters with the information they need to reach an informed decision.

Public health and safety

Advances in medical research pose ethical dilemmas, which engage researchers who develop the techniques and physicians who make clinical use of them. Newly developed drugs and medical technologies must be vetted by public agencies for safety and efficacy so as to protect the public health. Psycho-pharmaceutical agents are subject to regulation. Concern for the conquest of diseases once impervious to science requires decisions on an unprecedented scale in the allocation of public funds, for public health and medical treatment as well as research. In all these efforts, scientists engaged in biomedical research and application are called upon to offer their expertise and are subject to social regulation. Other safety issues in which scientists play an important role include the regulation of agricultural and industrial chemicals and nuclear power.

Innovation and economic development

Because technological innovation plays a vital role in economic growth and international trade, and especially in the technologically advanced countries, the strength of science and technology has become a critical national priority in every country. The unexpected consequences of some new technologies have also led to calls for advance 'technology assessment'. The exhaustion of fossil fuels has led to concern for the development of synthetic fuels and alternative, renewable sources of energy. The development of nuclear technology has led to controversy over safety and over the long-term disposal of nuclear wastes. Nuclear accidents, notably the disaster at Chernobyl in Ukraine, called attention to the need to monitor such activities and to their international and national dangers. On all these scores, scientists and technologists have been called upon for advice, for new research, and for the implementation of national policies.

The environment

Threats to the balances of nature, including species diversity, resulting from the use of chemicals, many of them of synthetic design, in agriculture and manufacturing, as well as from economic development, industrialization, and population increase, have engendered a movement for the protection of the environment. Environmentalism sometimes takes political form, and scientists in various disciplines have become prominent participants and even activists in various environmental causes.

There are numerous other areas of social interaction in which science and politics also converge. In forensic medicine, the increased reliance on DNA typing is perhaps the most dramatic example. Similarly, because an ability to appreciate medical and scientific testimony can be critical, the continued validity of the jury system has come into question, and the distinction between 'consensual' and maverick or 'junk' scientific testimony has also been raised. Space exploration is another area in which science and politics have become connected. The US–Soviet competition to reach the moon and explore outer space in the 1960s was sparked by the declaration of 1957 as an International Geophysical Year by the ICSU (McDougall, 1985: 118). Governments have also had to take steps to protect the integrity of the scientific process, as competitive pressures have led to an apparent increase in fraud among scientists. Campaigns for legislation protecting animal rights have led scientists to band together to defend the use of laboratory animals, but have also had the salutary effect of leading more humane treatment of animals used in research. While all these areas warrant inquiry, attention here will be restricted to the four areas in which scientists have been primarily engaged with government and public policy.

The incorporation of scientists and technologists into the political system

While the growing importance of science and technology has long been anticipated by visionary social theorists and novelists, the spectre of an impending technocratic 'Brave New World' has so far scarcely materialized (Huxley, 1932; Meynaud, 1968). On the whole, scientists have so far been incorporated into existing political systems. During World War II and the Cold War, some of those who worked in the American atomic bomb project acted out of political convictions to violate security regulations and betray atomic secrets to the Soviet Union. Others have become radical activists on environmental issues, challenging government policies by direct action. By far the majority of scientists act within the political process, whether as employees and advisers to government agencies or as citizens. They do not form a cohesive single elite with a uniform set of attitudes. They divide along the familiar left–right spectrum, with more theoretical scientists apt to be more on the left than those specializing in experimental research and industrial or agricultural development (Ladd and Lipset, 1972). Some see their political role as necessarily compromised by their dependence on public support (Mukerji, 1989).

Every government maintains a scientific advisory process through which scientific advice is solicited. In democracies, they are encouraged to engage in public debate, subject to concerns over the public dissemination of classified research. Some parliamentary cabinets include a portfolio for science and technology. In others, there are efforts both within the executive and the legislatures to coordinate science policy-making, to channel advice to politicians, and to provide research facilities for obtaining policy assessments (Golden, 1991). Every government provides research subvention, some through more homogenous and centrally controlled bodies, others in more pluralistic form, and scientists are compelled, no less than other interest groups, to fight for support and jockey for priority amongst themselves in setting formulas and allocating project support. 'Big science' projects such as particle accelerators, space stations, and thermonuclear energy projects inevitably become political issues, complete with pork-barrel patronage, both on the domestic and international level. Rather than replace the democratic process, the entry of scientists into the political arena has more often strengthened it. In the process of attempting to win support, arguing for or against particular technologies, and calling attention to the implications of innovation, scientists have helped revitalize and invigorate the adversary process on which parliamentary democracy greatly depends.

Scientists have also helped strengthen international cooperation. Their interest in international collaboration has led to the formation of many committees and institutions. Notable among them are the International Council of Scientific Unions (ICSU) and such United Nations bodies as the World Health Organization (WHO), the United Nations International Development Organization (UNIDO), the United Nations Educational, Scientific and Cultural Organization (UNESCO), and the recently formed Intergovernmental Panel on Climate Change. UNESCO has sponsored the International Center for Theoretical Physics; UNIDO has created International Centers for Genetic Engineering and Biotechnology and the International Center for Science, with several components. The World Bank supports thirteen institutes designed to introduce scientific agriculture in the developing countries (Golden, 1991: 69). Several nations (the US, France, and Britain) have collaborated to establish the International Thermonuclear Experimental Reactor project (ITER) to develop fusion power. The East–West conflict engendered the formation of the Pugwash movement. Within the European Union, several comprehensive scientific programs have been launched, including EURA-TOM for nuclear research, the Centre European pour la Recherche Nucleaire (CERN) for experimental high-energy physics established in Geneva in the 1950s, ESPRIT (European Strategic Program for Research and Development in Information Technology), BRITE/EURAM (Basic Research for Industrial Technologies in Europe/European Materials), RACE (R&D in Advanced Communications-Technologies in Europe, the all-purpose EUREKA (European Research Coordination Agency), the European Space Agency, the European Molecular Biology Organization, and smaller laboratory settings like the Joint European Torus (JET) for fusion research (Golden, 1991: 45–55; Sandholtz, 1992). The Organization for Economic Cooperation and Development (OECD), established in 1960, tracks and compares patterns of research support. The International Institute for Applied Systems Analysis fosters cooperation in development of statistical models relevant to innovation and the environment. As a result of these developments, coupled with the movement of researchers among laboratories and the greatly increased speed with which the results of research are now communicated, and a corresponding and even richer development of technology transfer via multinational corporations of virtually global character, a thick and interactive network of scientific cooperation has transformed the traditional cosmopolitanism of science and its 'invisible colleges' into a very real form of world federalism often in advance and even at variance with the anachronistic division of the world into competing sovereign states.

While the dependence of scientists on public subvention inevitably compromises their complete autonomy and political independence, it does not altogether prevent them, especially in democratic settings, from venting their views on public issues in which their expertise is relevant. In some cases, scientists spark political action by revealing their findings. Prominent scientists and spokesmen for scientific organizations influence public opinion through the mass media and are often called upon to testify before public commissions and legislative committees. Those employed in private industry are more constrained than those who work in university settings and may need to break ranks and become 'whistle blowers' in order to call public attention to dangerous activities and cover-ups. Some institutions, like national academies of science or international scientific unions and disciplinary organizations, allow scientists to address public policy issues from a non-partisan platform, though such organizations often shy away from controversial issues. Other organizations are formed deliberately to enable scientists of particular persuasions to take positions on public issues. And in other cases, notably in connection with the environment, scientists cooperate with others. The result is a very mixed picture. In some matters, scientists present a united front, in others they divide along institutional, disciplinary, and ideological lines. To explore the question in order to provide some coherence, this chapter will first examine the various dimensions of the links which have been discerned between scientists and politics and then propose some generalizations that can be drawn from this welter of experience.

Arms and the scientist

The involvement of scientists with warfare goes back sporadically many centuries, but became a pressing matter only in recent decades. Archimedes designed huge engines of war which enabled his native Syracuse to repel a Roman invasion in the Third Century BC. Galileo worked in the Venetian arsenal and was awarded a thousand ducats and a tenured professorship by the Doge for inventing a militarily useful telescope. Da Vinci suppressed a design for a submarine 'on account of the evil nature of men' (Brodie and Fawn, 1962; Morton, 1969). The use of interchangeable parts introduced by the American inventor Eli Whitney greatly improved the manufacture of rifles and other guns. The development of gunpowder, ballistic missiles, the railroad, the telegraph, and the steam engine (especially in the form of the railroad engine and the steamship) had important implications for warfare in early modern times. It was during World War I that the potential applications of science to warfare became a matter for government-sponsored projects. Because of the development of new explosives and poison gas, this war became known as the 'chemists' war'. To some extent, it might also have been considered the civil, mechanical, and aeronautical engineers' war, in view of the development of the automobile, the armoured vehicle, and the aeroplane, but these did not yet play the role they were shortly to assume in making warfare mobile. World War I was still fought overwhelmingly on classic lines of trench warfare.

The great watershed between old hand-to-hand combat and modern warfare was World War II. Culminating in the development of the atomic bomb, by the use of which the war was brought to an end, World War II saw a mechanization of warfare on a massive scale, one that obliterated the old distinction between the battlefront and the home front and brought on the 'Century of total war' (Aron, 1954) as 'strategic bombing' became a euphemism for the destruction of cities and their inhabitants. Such inventions as radar (invented by Sir Robert Watson-Watt) and sonar became critical devices for defence against attack by air and sea respectively, and radio communication, coupled with the sophistication of armoured vehicles and the extended range of artillery, enabled military commanders to adopt highly mobile forms of tactics and strategy. Toward the end of the war, the Germans introduced jet aircraft and the guided missile, developed by a team headed by Werner von Braun, too late to have a major effect on the outcome, but very much harbingers, along with the atomic bomb, of the shape of warfare to come. In the process, scientists and engineers became major contributors to warfare (Clark, 1965; Snow, 1961), although their role in the war itself was still ancillary in the sense that they served on committees and took charge of projects but were not consulted on the use of the technologies. War may already have become too important to be left to the generals, in Clemenceau's famous phrase, but during World War II the generals, not the scientists, were exclusively in charge.

Several refugee scientists – Albert Einstein, Leo Szilard, Edward Teller, and Eugene Wigner among them – helped initiate the American 'Manhattan Project' to develop an atomic bomb by calling the attention of President Franklin Roosevelt to Enrico Fermi's work and warning that the German researchers were undoubtedly working to achieve the bomb. On the German side, such leading physicists as Walther Gerlach, Werner Heisenberg, Otto Hahn, and Carl von Weizsäker were consulted, but the project was not given the priority it would have needed to become successful, in the expectation that it would take too long to be of use in the war and would divert scarce resources. In one of the stranger episodes of the war, Heisenberg visited occupied Copenhagen in 1941 and spoke guardedly to Niels Bohr, the most revered European physicist. Bohr interpreted his elliptical remarks as an effort to persuade him to agree that scientists on both sides refrain from developing the atomic bomb (Heisenberg, 1971). Alarmed, Bohr warned the allies that the Germans might well be trying to develop the bomb. In fact, the Germans made little progress in their atomic bomb project, largely because it was not thought likely to be successful in time to affect the war and was therefore given low priority (Beyerchen, 1985; Irving, 1967; Rhodes, 1986). Aware only that German physicists were highly capable and as knowledgeable as they about recent developments, the scientists who worked on the Manhattan Project, especially the refugees, worked as hard as they did out of fear that Germany might win the race. Japanese physicists also worked on an atomic bomb project, but shortages and working conditions made difficult by allied bombardment, inhibited their progress. The Japanese also experimented with biological warfare, using disease-causing agents against the Chinese.

When the war ended with the dropping of two atomic bombs on Japan, many of the researchers who had taken part in the project chafed at being excluded from the political decision to use the bomb against civilian targets. While most adhered to the traditional notion that science is politically neutral, some had already been politicized by developments in the inter-war period. The success of the Bolshevik revolution and the struggle against fascism in the 1930s and 1940s led many scientists, like other intellectuals, to become anti-fascist and to see their work as vital to the political struggle (Bernal, 1939). A few who were involved in the atomic bomb project (notably the German-born theoretical physicist Klaus Fuchs (Moorehead, 1952) and the American, Theodore Hall), betrayed secrets to the Soviet Union, rationalizing their violation of security regulations on the ground that they were helping an ally in the war against fascism, that they were serving the cause of Communism, and that they were helping the cause of peace by denying the United States and Britain a monopoly over atomic weapons. In the final months of the war, a number of scientists at work on the Manhattan Project sought to prevent the military use of the bomb but were unsuccessful. Afterwards, they formed the nucleus of an 'atomic scientists movement', seeking to alert the public to the need for international control of dangerous atomic energy research (Smith, 1965). In 1946 they formed the Federation of American Scientists and campaigned to make clear that because the 'secret' of the atomic bomb cannot be monopolized and because there could be no effective civil defense against atomic attack, only international agreement could avert a nuclear catastrophe.

In the early 1950s, controversy broke out in the United States over the adherence of scientists to the loyalty and security regulations. This problem was especially acute in the early years of the Cold War due to virtual hysteria over national security and subversion in the United States, a state of mind reinforced by revelations that there had indeed been espionage at Los Alamos. The withdrawal of the security clearance of J. Robert Oppenheimer, the physicist who had been the director of the wartime project, became a cause celebre (Alsop and Alsop, 1954; Dupré and Lakoff, 1962; Lakoff, 1966; Stern with Green, 1969; United States Atomic Energy Commission,

1954). The affair began because Oppenheimer had opposed a crash program to develop a thermonuclear bomb. The most extreme of his critics saw this as part of a pattern of disloyal behavior, which made him a security risk. The withdrawal was upheld on appeal, not on grounds of his opposition to the crash program (in the absence of which, however, there would have been little reason to review his clearance) but because he had shown 'defects of character' in temporarily misleading security officers to protect a friend he thought innocent.

Other scientists also played a role in events connected to nuclear weapons. Edward Teller, who criticized Oppenheimer as a security risk, became a vigorous proponent of the development of the thermonuclear bomb and of civil defence (Rhodes, 1995; York, 1976). Linus Pauling, twice a Nobel Prize winner, led a crusade to stop nuclear testing because of the harm done by radioactive fallout. Eventually this campaign succeeded partially with the signing of the limited test ban treaty of 1960, whereby at least the two major nuclear powers agreed to confine their tests to underground chambers. The Polish-born physicist Joseph Rotblat organized the Pugwash movement, which served as a channel for discussions between scientists on both sides of the Cold War, eventually winning the Nobel Prize for Peace for his efforts (Rotblat, 1972). Andrei Sakharov, the father of the Soviet thermonuclear bomb, sought to promote arms control and was exiled to Gorky for making his views public and for calling for liberalization of the Soviet system, and was also awarded the Nobel Prize for Peace (Krementsov, 1997). Herbert F. York, director of the Livermore Laboratory at which the American thermonuclear bomb was developed, became an advocate of arms control (seeing the arms race as 'a race to oblivion') and ambassador to the comprehensive test ban treaty talks and a participant in these talks (York, 1970, 1987). Others who took an active part in public debates on military policy included the physicists Sidney Drell, Richard Garwin, and Hans Bethe, and the engineer Jerome Wiesner, who became outspoken opponents of the anti-ballistic missile (ABM) program, which they thought would be ineffective and would only stimulate a Soviet offensive response (Cahn, 1971). Thousands of scientists and engineers in all the advanced countries worked on military projects, most with a sense that they had no special right to more of a say than other citizens in what was done with their work. Some campaigned to force universities to divest military research laboratories (Nelkin, 1972). Some became military strategists, though before long a special new breed of strategists, some relying on systems analysis and based in organizations like the RAND Corporation, challenged their qualifications to pronounce on strategic questions. The mathematicians John von Neumann and Oskar Morganstern popularized the use of game theory in strategy, and a combined community of academics developed the new language of nuclear strategy, introducing such terms as mutual assured deterrence, counterforce/countervalue, distinctions between first strikes and retaliatory strikes, and such euphemisms for the destruction of civil targets as 'collateral damage' (Freedman, 1981; Herken, 1985; Kaplan, 1983; Smith, 1966; Wohlstetter, 1964).

Other independent researchers made effective criticism of the reliance on various military technologies. Theodore A. Postol revealed after the 1995 Gulf War that the Patriot anti-missile defence system had been ineffective, contrary to published claims by the Defense Department. This assessment was gradually accepted by impartial analysts, but the Raytheon Company, which produced the Patriot missile, withdrew a planned contribution to NET out of displeasure with Postol's public utterances and the university's unwillingness to interfere with his academic freedom. Another independent researcher, Kosta Tsipis, a physicist at the Massachusetts Institute of Technology, sought to deflate claims for various weapons systems by making his own analyses of Pentagon weapons planning without access to classified information.

Scientists and engineers have not confined their role in war research to the advanced countries. Several developing countries have acquired nuclear weapons, including India, Israel, and Pakistan. Others, including Iraq under Saddam Hussein and Libya under Mu'ammar al-Qaddafi, have sought to acquire nuclear weapons as well as chemical and biological weapons ('the poor man's atomic bomb'). With the significant exception of Sakharov's efforts in the Soviet Union, which were rebuffed by the leadership and unknown to the general public, scientists have been most active politically in democracies. Only late in the history of the Soviet Union were scientists like Evygeny Velikhov and Roald Sagdeyev able to exert an influence. In all cases, scientists like other experts, have been held 'on tap but not on top'. They have nevertheless played a unique role as a catalyst in public debates and in movements for arms control. It is now common for leading scientists to be asked for their views on public policy in areas of their expertise, but to some extent institutionalization has replaced the emphasis on the opinions of leading figures in earlier times. Now it is more likely to be the national academies and professional organizations of scientists which are consulted for opinions or asked to conduct policy-oriented studies.

Health and safety

Although much research dealing with pharmaceuticals is conducted by private companies, government support has become important for basic research in health-related science. Most countries also maintain national health services or health insurance systems in which physicians and biomedical researchers are actively engaged in the making and implementation of public policy. Medical scientists administering research support systems (like the American National Institutes of Health or the British Medical Research Council), advocate policies for the support of health care and medical research, reviewing proposals for new drugs, and carry out research. Organizations of medical scientists contribute to policy debates (such as those over the patenting of new life forms and research into recombinant DNA) and one organization, Physicians for Social Responsibility, interests itself in issues that go beyond a narrow definition of health to include nuclear war and the environment. The World Health Organization, a UN agency, involves public health specialists in matters of global concern. Another independent group, Médecins sans frontières, assists people in developing countries who are in dire need. Health and safety issues connected with nuclear power have been a particular concern of the American Union of Concerned Scientists.

Apart from more general issues of health care, the key policy debates in which medical scientists have been engaged have been the debate over recombinant DNA research and the need for public campaigns to eradicate various diseases (notably cancer, heart disease, AIDS, and various tropical diseases). The recombinant DNA debate concerned the safety of experiments to design new life forms. At first, it was thought these new forms might not be easily contained in laboratories and could escape to cause mutations and spread infections. The American 'war on cancer' was an example, however, of how non-scientific elites could gain passage of legislation over the objections of qualified researchers, in this case both of the National Institutes of Health and most of the biomedical scientific community (Rettig, 1977).

Scientists have also served on advisory panels to review drug proposals and to evaluate the risks posed by chemicals. In these processes they have often disagreed in making professional assessments. These disagreements sometimes reflect differing institutional commitments: scientists affiliated with industry are more likely to oppose regulation than those associated with environmental organizations. 'Regulatory science' has been shown to be less exact than ordinary science and therefore more susceptible to technical bias and political judgement (Jasanoff, 1990).

The backlash against technology and science has led to efforts to prevent experiments with cloning and other forms of genetic intervention, including the use of gene modification in livestock and fruits and vegetables. Anti-vivisectionists have succeeded in forcing improvements in the treatment of laboratory animals and restrictions on the experimental use of animals where other alternatives are available, but more radical elements have disrupted experiments and threatened the lives of scientists conducting them, and are demanding a total ban on the use of animals in experimentation. Adherents of various European Green parties have urged the European Union to block importation of American products using gene modification or at least to require that the products be labeled to differentiate them from conventionally produced versions. In 1997, for example, a proposal was put forward in Switzerland, with the support of the Swiss Green movement and the Swiss Socialist Party, which would have resulted in a constitutional prohibition of the use and patenting of gene-modified animals, even including flies and worms, and the cultivation of gene-modified plants. Since Swiss researchers and companies are among the leaders in pharmaceuticals and biotechnology in general, such a ban would pose a sweeping obstacle to work in this field and possibly set a precedent for prohibitions elsewhere. Both houses of the Swiss parliament declared their opposition to the measure, however. Such campaigns have led some analysts to suggest that some elements of the Western electorate have adopted a new set of 'post-materialist' attitudes in reaction to the growing reliance on technology (Poguntke, 1993). The failure of the German Greens to maintain their early electoral success, however, suggests that efforts to exploit these attitudinal changes to alter the political landscape must overcome the more traditional difficulties of factional conflict, the need for pragmatic compromise, and the countervailing appeal of economic interest (Scharf, 1994: 5).

Innovation

Scientists and technologists are part of a complex and not fully understood process whereby economic innovation takes place. They are by no means the only important actors. States and entrepreneurs, sources of investment capital, educational facilities, and the stimulus of public demand are among the forces at play (de la Mothe and Paquet, 1996a, 1998, 1999; Freeman, 1990; Smith and Barfield, 1996). Sometimes governments established cooperative efforts to promote innovation. In the US, the federal government established a consortium of high-technology companies as SEMATECH. A joint public private enterprise created the cooperative Space Program. Military research support with civil 'spin off' has nurtured major industries, including computers, aircraft, and electronics. In Europe, inter-governmental cooperation has underwritten such efforts as the Concord supersonic aircraft and the Airbus civil airliner. Many scientists and technologists work on projects either designed to have direct economic benefits or that turn out to have such benefits. When they work for profit-making companies they may well be constrained to present the point of view of the company. They become advocates for particular technological advances. In other cases, scientists work in non-profit settings in which they are more insulated from economic pressures and are sometimes more likely to become critical of economic innovation. Those employed by auto companies are often in a defensive posture in resisting calls for pollution filters, increased gas mileage or the use of safety devices like airbags which would add to the cost of the automobiles. It is often those outside the industry itself who campaign for innovations in design or in the weight of cars that promote efficiency (Von Hippel, 1991: 173–90), but the companies themselves, subject to government regulation, must actually develop and introduce the technologies. Those researchers who warned of the health-related dangers of cigarette smoking were not employees or contractors of the tobacco companies but members of medical facilities and university research centers. Those in favour of various forms of nuclear power are likely to be employed in the nuclear power industry or in the fields of research related to the industrial applications. 'Public interest' science organizations attract researchers anxious to display social responsibility and are often critical of those who work for these profit-making enterprises. A serious problem in achieving effective regulation, in areas like nuclear energy, is that often the only people capable of doing the regulating are those with experience in the affected industry.

The need for technological innovation for economic growth raises special problems in developing countries. Some scientists work with international organizations committed to helping the developing countries achieve 'appropriate technology' but others are likely to affect the developing countries indirectly by working for large multinational enterprises, in major industries, which have an effect on them. Norman Borlaug's 'Green Revolution' has enormously increased yields for rice and wheat in the developing countries of Asia and in Mexico, though it has been criticized for increasing the hold of large-scale owners in these countries and threatening species diversity (Brown, 1970).

Environment

In America, the movement for conservation was sparked by the early 'eco-pioneers' such as Henry David Thoreau, Aldo Leopold, and John Muir. The environmental movement was stimulated by such prominent life scientists as Rachel Carson, Barry Commoner, and Rene Dubos (Carson, 1962; Commoner, 1971; Shabecoff, 1993). Rachel Carson warned against the damaging effects on animal life of DDT. Carson's warnings prompted inquiries and regulatory action, including the banning of DDT and other pesticides, although in retrospect certain of her direst warnings, such as that the robin would become extinct in North America, did not come to pass

(Easterbrook, 1995: 80). Commoner led in organizing the first Earth Day in 1970 and a large environmentalist movement that followed from it. Dubos issued what became a kind of slogan for the movement: 'Think globally, act locally'. Others have also played important roles in arousing public opinion on specific issues (Vig and Kraft, 1994). The entomologist and socio-biologist E. O. Wilson has stressed the importance of species diversity. Matthew Meselson spearheaded the inquiry of the American Association for the Advancement of Science into the health effects of chemical defoliants used by American forces in South Vietnam. Charles Wurster's research on the effects of DDT on birds led to the founding of the Environmental Defense Fund, which undertook campaigns of litigation to regulate and ban the use of pesticides. Garrett Hardin wrote a celebrated essay on the 'tragedy of the commons', in which he warned that failure to protect the global commons could reproduce an earlier folly on a far greater scale. Paul Ehrlich's book predicted a catastrophic 'population explosion'. Although it did not materialize, his warnings helped galvanize the environmental movement. The naturalist Fairfield Osborn emphasized the loss of habit and wildlife (Shabecoff, 1993). Catastrophes such as the release of noxious gas by a chemical factory in Bhopal, India caused great suffering and energized environmental activism in the developing countries. Large numbers have taken part in broad membership environmental organizations like the American Sierra Club and Natural Resources Defense Council and some have enlisted in more radical organizations like Earth First and Greenpeace, which employ direct-action techniques to block or interfere with projects of which they disapprove. Environmental organizations frequently use legal challenges to prevent polluting activities. The Sierra Club maintains a legal defense fund, which at any given time is likely to be engaged in many such lawsuits. Laws making polluters liable opened the floodgates to litigation and inflicted great costs on corporations found guilty of pollution. Similar lawsuits have been launched against the tobacco companies by state attorneys general for knowingly causing illness by selling their products. In a number of countries Green Parties have been formed which work closely with environmental scientists but are generally leery of other types of scientific work. The more radical environmental organizations tend to attract only politically radical scientists, except insofar as they reflect a division within the scientific community between life scientists and physical scientists. Some life scientists tend to oppose the development of chemicals and forms of energy (notably nuclear power) which they see as environmentally detrimental. The physical scientists and engineers involved in such work usually find themselves in the opposite camp, defending the morality and safety of their work. Some radical 'enviros' believe in the 'Deep Ecology' of the Norwegian philosopher Arne Naess, which contends that non-human life has a value independent of that of human life and therefore that humans have no right to reduce natural diversity or degrade the environment beyond what is necessary to sustain human life.

In the 1970s physicists and others joined the effort to block development of the supersonic transport plane on grounds of its environmental impact in injecting large amounts of water vapour into the upper atmosphere and causing sonic boom (Primack and von Hippel, 1974). Worldwide efforts have been made to protect endangered species, such as whales and rhinoceros. Passage of the Superfund legislation was an outgrowth of strong environmental protests over the dumping of toxic chemicals and the shock caused by reports of contamination in particular places, like Love Canal near Buffalo, NY.

In the 1990s, scientists have become especially active in warning about the dangers to the global environment. The chemists Mario Molina and Sherwood Rowland, of the University of California, Irvine, called attention to the depletion of ozone in the atmosphere leading to an unprecedented international agreement to eliminate chlorofluorcarbons. In 1997, over 2,000 climate scientists organized in the Intergovernmental Panel on Climate Change warned that because of 'anthropogenic effects' the release of CO^2 into the atmosphere from the burning of coal, oil, and natural gas, the earth could experience 'global warming' due to the enhanced greenhouse effect, with results that could be catastrophic, especially for poorer countries. Virologists have joined with climatologists, warning that as heat rises, hundreds of thousands of deaths might result from extremes of floods and droughts, storms and heat waves. Although the scientists who think that global warming is a serious concern readily admit that the supporting data are not yet beyond doubt, and that those in advanced countries, with vastly superior systems of physical control and public health, will be better able to protect themselves from harmful consequences, they content that protective measures are already warranted. A minority of scientists dissents from this view, arguing that the evidence is far from sufficient to warrant the costly efforts that would be required to cut back global emissions due to the burning of fossil fuels. Economists have joined the effort to argue in favour of a market in emissions trading. International agreement has been reached on vague guidelines and an effort is being made to impose limitations on emissions in the advanced countries. Scientific judgement is playing a major role in this effort.

Scientists and politics: assessing the changes

Although the interrelationship between science and politics has not altered government in the way that the visionary prophets of technocracy anticipated, it has changed the processes of government by compelling the adoption of more sophisticated, often more quantitative techniques, including systems analysis, cost-benefit analysis, and risk analysis. As scientists battled over such questions as the validity of proposals for an anti-ballistic missile system, they developed competing estimates of the technical feasibility of such systems, citing known data and physical laws. In debating the practicality of test ban treaties, they had to estimate the possibility of cheating and the likelihood that incidents of cheating could be detected. The systems analysts sought to replace 'seat of the pants' judgments by experienced soldiers with presumably more objective studies. Computer simulation and calculation enabled a dramatic increase in sophistication in weapons design and impact analysis. In the case of environmentalism, the early efforts were usually crude because certain chemicals were identified as hazardous and likely to be banned entirely, as by the Delaney Clause of the Food and Drug Act in the United States, which banned all potential carcinogens regardless of dosage. In later years, as measurement devices improved, it became possible to argue that minute concentrations were not likely to be harmful. A conflict has arisen over the strict environmentalist idea that any dose is dangerous and the economic argument that one should rely on dose-response ratios to determine when a substance is dangerous and when it is not (Wildavsky, 1995: 5, *passim*). General Circulation Models (GCMs) relied on by the Intergovernmental Panel on Climate Change, and therefore on the governments negotiating to prevent dangerous global warming, suffer the difficulties attendant on all complex models. They are reductionist because they are unable to contain all the data necessary to have a complete picture of something as complex as climate change. They require compromises or 'tuning' and therefore may be inaccurate.

Government agencies tend to be suspicious of outside advice, regarding this as likely to be biased and lacking in authority. They prefer their own 'in-house sources of information'. But extramural advice enters the policy debate, whether solicited or not, and plays a useful role in providing independent perspectives. Scientific advisory bodies located in the executive have the benefit of providing easy access to the agencies with the highest authority, but they are in a compromising position because they must conform their views to those of the administration (and may often be guided by political appointees chosen for their adherence to that point of view). When they do not, they run into difficulty, as happened in the United States when the President's Science Advisory Committee opposed the administration's views on several issues and was abolished by President Nixon. The fact that it was subsequently reconstituted, as the White House Council for Science and Technology, suggests a recognition within the American executive of the value of scientific advisors, but its political structure makes it obvious that there is an inherent problem in creating such a mechanism and making it truly representative of the varied voices within scientific communities. It is important, to avoid the danger of suppression and bias, that there be multiple sources of policy study and advice. The decision by President Ronald Reagan to embark on the 'Strategic Defense Initiative' (SDI), an effort to build an impenetrable defense against nuclear attack, was made in consultation only with sympathetic scientists; the great majority of researchers considered it beyond the scope of practical feasibility for the foreseeable future (Lakoff and York, 1989). The project was soon cut back and reduced to more modest goals when its technical goals proved unrealistic.

Scientific organizations have made a number of important efforts to strengthen communication between scientists and politicians. In the United States, the Congressional Research Service (CRS) serves legislators as a research arm on technical issues. Another important effort to improve legislative understanding of scientific issues was made jointly by the American Physical Society and the American Association for the Advancement of Science in setting up the Congressional Fellows Program in 1973. Under this program, young scientists are attached to a congressional office or committee. Similar programs have been established by other government agencies. The United States also created an innovative Office of Technology Assessment, which provided its services to members of the Congress but was terminated in a cost-cutting crusade early in the 1990s. At about the same time, the same fate befell the Science Council of Canada, an agency which also carried out similarly valuable studies designed to improve the making of science policy. There is considerable variation, however, in the relationships between scientists and the state. Solingen has suggested four 'ideal types' to account for the variability: happy convergence, passive resistance, ritual confrontation, and deadly encounters (Solingen, 1994).

In developing countries, even when they are in harmony with the state's goals, scientists face a special problem because they cannot be sure that state policy will be sustained. In Mexico, it is said that science and technology have been 'pasteurized' because they are periodically warmed up and then cooled (Golden, 1991: 308). The lack of strong educational infrastructure means that the best scientists are apt to be demoralized, and unable to work at the frontiers of research in their disciplines. They can find great rewards and encouragement, but not the stimulus of the work environment they can expect in the more advanced countries. Only if centers of excellence are created and linked to centers in the advanced countries will this difficulty be overcome (as it is in some countries such as Israel and Singapore).

In general, scientific communities are engaged in politics on a number of fronts: as supplicants for public support, as advisers to government agencies, and as policy advocates both within the councils of state and in the public forum. Their participation in decision-making has become indispensable for rational policy-making, especially inasmuch as great sums of money are often at stake, either in making commitments to research or in foregoing or regulating technological activities. Whether it makes sense for governments to invest in long-term projects to advance basic knowledge by building ever larger particle accelerators or sending missions into outer space, or to undertake more practical efforts, to develop thermonuclear energy and other renewable sources of energy, are questions in which scientific judgment is indispensable, even though it is apt to be biased when those whose professional careers are invested in these projects are asked to give their opinions on their desirability. Often there is a moral dimension as well as an economic one in these matters. The advent of nuclear weapons forced a reconsideration of the thinking behind the concept of just war. (Is it morally acceptable to threaten nuclear retaliation against civil populations for the sake of deterrence?) Developments in biology and medicine opened more conundrums. (Should research on recombinant DNA be permitted even if there was a risk that new forms of life might be released that would cause harmful mutations? Should human life be cloned? Should fertility drugs be freely prescribed or regulated? How should scarce medical technologies be rationed? Now that medical technology can prolong the 'life' of hopelessly ill and comatose patients, should society recognize, along with the right to life and liberty, a right to die with dignity? Who should make these decisions?) A new profession of bioethicists has been developed to serve as a mediator between society and physicians, but no more than scientists can bioethicists claim the right or basis for judgment to make such decisions for others. Many worry that the progress of science has created a 'technological imperative', according to which, whatever can be done, will be done, somehow, by someone. To preserve human autonomy, it is sometimes said, it may be necessary to limit the activities of scientists and science for the sake of preserving a sense of human control.

Such difficult questions can be addressed only with the help of informed understanding. Scientists must therefore play an active role in the political process and ordinary citizens must be well enough educated to keep abreast of developments in science and understand the complex issues that science presents. The modern age is, almost by definition, an age in which the crucial social questions are often defined by the progress of knowledge. Science is more than a technique for achieving understanding or control, however; it is a way of understanding. It does not, however, contain explicit ethical or political guidance, except insofar as it makes its pursuers aware of the ultimate mystery of creation and reminds them of the dangers of hubris. Politicization need not require the breaking of ties among scientists or the adoption of chauvinistic forms of nationalism. The 'republic of science' remains a transnational community, with its own 'invisible colleges', its own norm of openness. 'Science is by its nature universal: the truths scientists pursue are not national truths; they are the same everywhere, unanimously acknowledged' (Salomon, 1970: 315). Science remains 'the pre-eminent transnational community' (Rhodes, 1986: 788). As such, it may well exert a benign influence by fostering a sense of common humanity at a time of increasing political and economic interdependence.

CHAPTER THREE

Governance and the New Production of Knowledge

Michael Gibbons

Introduction

It may be, as John de la Mothe has observed (de la Mothe and Dufour 1995), that the practice of science and technology policy in some countries – classically defined in strict terms of funding for research – is in the doldrums; that is, it is in a state of intellectual depression or lack of fresh air. This may be due to the perennial tension between supply and demand; between the institutions which conduct research and governments which try to develop policies to meet social needs. But as de la Mothe and Paquet have subsequently noted (1995, 1996), to formulate the practical or operational issue in this way, however, is to adopt a view – as has been adopted by several governments – in which scientific research and the policy-making processes are seen as being fundamentally separate or opposed. In this operational view, the processes that lead to the production of new knowledge – whether in scientific or technological domains – are working adequately. The difficulty is to harness, in increasingly effective ways, these outputs to socially useful purposes and this is what policy-makers are continuously attempting to do. Thinking on science policy has, by now, gone through at least three phases of development, but in no phase has the possibility that the knowledge production process changed radically ever been considered. Instead, the view that knowledge production is essentially a distinct set of activities from the policy process has continued to predominate. As a result, most of the effort has gone into making the linkages between supply and demand more effective. Despite working with a deficient model, science policy has none the less made a difference. For example, the way in which research problems are identified and funded is now a much more complex process than was envisaged by Weinberg in his seminal papers on the criteria for scientific choice. To have modified what the scientific community regard as important scientific problems is no mean achievement. Yet as policy thinking moves beyond production, the distribution of knowledge assumes greater importance. As will be shown below, these issues grow out of a transformation of the knowledge production process itself, and call into question many of the assumptions that have guided thinking in the earlier phases. For this reason, it will be useful to set the stage by referring to some of the characteristics of earlier science policy thinking.

Policy for science

In the first phase, the problem as expressed in the writings of Vannevar Bush (Bush, 1946) and Alvin Weinberg (Weinberg, 1963) among others, was posed in terms of working out a policy for science. The main issue, then, was the growth of the scientific enterprise *per se*. The key questions were concerned with criteria for choice within science; setting up guidelines for choosing between expensive projects, often in different disciplines. This vision of science policy, in which the key decisions were to be taken by scientists, now seems untenable if not naïve. None the less, it still lingers in the minds of many in academia as the norm of a proper policy for science. Such a policy, however effective it may have been, has now become inadequate. And that is because it addresses itself principally to what is happening within disciplines whereas the global dynamics of knowledge production have become much more concerned with what is happening outside or alongside them. So much is happening outside the traditional disciplines that it seems folly to formulate policy entirely from within them.

Science in policy

In the second phase, both scientists and policy-makers advocated a reform: policy needed to shift from 'policy for science' to 'policy in which science was seen to support the objectives of other policies'; a shift to science in policy. The Brooks Report (OECD, 1971), the Rothschild Report in the UK (HMSO, 1972), the Research Applied to National Needs (RANN) programme in the US, were examples of this new perspective. The intention was that science and technology should play a key role in achieving the diverse policy objectives of a modern industrial state rather than simply aiming at the development of science itself. Yet, in neither of these first two phases was much attention given to how science might contribute to national well-being. That there were potential benefits to be had from science was unquestioned, but it was not the scientist's job to extract them. In the event, deteriorating economic performance throughout the late 1970s and the early 1980s in virtually all industrial economies eventually forced a critical reappraisal of the notion of science as the locomotive of economic performance. This in turn, brought forth another policy shift.

Policy for technological innovation

During the 1980s, declining economic performance and increasing world-wide competition forced policy-makers to narrow their perspective on the role of science in achieving national goals to the single question of how to hitch the scientific enterprise to industrial innovation and competitiveness. By means of a new range of initiatives aimed at promoting, first, strategic and, then, generic technologies policies shifted to technology as a more effective base from which to support national industry. In part, this was a response to falling competitiveness vis-à-vis Japan, but it also reflected the widespread belief that, at root, the technological base of the economy was depleted. It then became one of the aims of policy to repair these fundamental structural weaknesses by supporting the development of infrastructure technologies (semiconductors, new materials, etc.). A strengthening of those technologies which underlay industrial competitiveness was felt to be needed rather than stimulating innovation through specific product and process developments. This change of orientation and belief clearly exhibits some of the attributes of knowledge production in what we call Mode 2: a blurring of the distinction between science and technology, the creation of national (Alvey in the UK and ICOT in Japan), and in some cases supranational (ESPRIT and EUREKA) programs in particular technological regimes built around configuring national resources, the establishment of networks and other informal modes of communication amongst active partners, growing familiarity of university scientists with working in large, often multinational, teams.

At the moment, the impetus of these policies seems to be slowing despite, or perhaps because of, the fact that in many countries, particularly in the US, productivity and, hence international competitiveness has still not improved substantially. Throughout this period, the scientific communities in many countries have stressed the importance of basic science to industrial well-being. Industry, in its turn, under pressure of rising costs, has had to narrow its commitments to basic research. Currently, policy is log-jammed trying to maintain a creative tension between a vigorous scientific enterprise and the imperatives of competitive industrial structure. It is contended that science and technology policy has been put into this quandary because the goals of policy have been broadened without questioning the fundamental presuppositions it entertained from the beginning. The question that has not yet been fully tackled concerns the contribution to economic performance that can be realistically expected from disciplinary-based sciences, institutionalized largely in universities, and driven, intellectually, by internal considerations. These problems are now recognized but have not been solved because the underlying assumptions about the role of science in the economy have yet to be critically addressed.

Under these new conditions, science and technology policies (phases 1 and 2) and innovation policies (phase 3) can no longer be regarded as functionally separate. Indeed, this is already the case in many countries where, under the label, 'science and technology policy' or research policy, it is actually 'innovation policy' which is being pursued. Generally, this is done with only meagre success because the presuppositions of older science policy thinking linger on and continue to structure the thinking of policy analysts and decision-makers. In the new phase, innovation policy, if it is to be efficient, will supplant the older science and technology policy thinking. It will be a new type of innovation policy, predicated upon a broader understanding of the innovation process and of the constitutive role of knowledge and the knowledge producing institutions in it. A key element in this new understanding is already becoming clear. It is that people in their fungibility, multicompetence and capacity to connect with others are the crucial resource.

Policy problems cannot be properly addressed until policy-makers take into account the many significant changes taking place in the production of knowledge, in industry as well as in the traditional sites where science is practiced. The traditional approach has been to export the problem of deriving economic benefits from science and technology to the people who manage the interface between science and industry, leaving activities on either side of the interface largely unchanged.

Ironically, this has occurred when many of these interfaces have become more permeable. The permeability has been brought about not by policy but because the best scientists realize they need to interact more strongly with knowledge created outside.

In what follows it is suggested that the current mode of knowledge production – the way that research is carried out at the leading edge of the most advanced studies whether in science, social science or the humanities – is changing dramatically. Further, it is contended that this development which is being led as much by the internal dynamics of knowledge production as by government policy or by increasing competitiveness has profound implications for the future shape and social function of all the institutions of science.

The disciplinary structure of science – Mode 1

The research structures that have gradually been put in place to guide scientific research not only in universities but in most institutions which purport to carry out research are supported by a set of research practices which ensures that results are sound. These research practices set the terms of what shall count as a contribution to knowledge, who shall be allowed to participate in its production, and how accreditation shall be organized. Together, these practices have generated what we know as the disciplinary structure of science and this structure, in turn, has come to play a central role in management and organization of research today. Of particular importance for what follows is the fact that the disciplinary structure is specialist. Whether in sciences, the social sciences, or the humanities specialism has been seen as a secure way to advance knowledge and its organizational imperatives have everywhere accompanied its implementation.

The disciplinary structure is also an organizing principle for teaching in universities. It provides a framework for the content of undergraduate curricula. The disciplinary structure is the essential link which connects teaching and research and which underpins the argument that in universities they properly belong together. Of course, research not only adds to the stock of specialist knowledge but transforms it as well. The research enterprise is a dynamic one. Its research practices articulate the disciplinary structure and, overtime, modify what are regarded as the essential ideas, techniques and methods that students need to be taught.

Most of the institutions of science, whether universities or research establishments, are built upon a model of knowledge production that has a disciplinary basis. This structure provides the guidelines about what the important problems are, how they should be tackled, who should tackle them, and what should be regarded as a contribution to the field. In brief, the disciplinary structure defines what shall count as 'good science'. This form of knowledge production is, for the purposes of this paper, labeled Mode 1. But a new mode of knowledge production may be emerging. The new mode of research production is appearing across the board in the sciences, the social sciences and the humanities. It is labeled Mode 2 and comprises a different set of research practices.

Changing research practices: Mode 1 and Mode 2

As pointed out in the previous paragraph, the contrast is with Mode 1, a term which refers to the form of knowledge production – a complex of ideas, methods, values, norms – that has grown up to control the diffusion of the Newtonian model to more and more fields of enquiry and ensure its compliance with what is considered sound scientific practice. For the purposes of this chapter, Mode 1 is meant to summarize in a single phrase the cognitive and social norms, which must be followed in the production, legitimation and diffusion of knowledge of this kind. For many, Mode 1 is identical with what is meant by science. Its cognitive and social norms determine what shall count as significant problems, who shall be allowed to practice science and what constitutes good science. Forms of practice, which adhere to these rules, are by definition 'scientific' while those that violate them are not. It is partly for these reasons that whereas in Mode 1 it is conventional to speak of science and scientists, it has been necessary to use the more general terms knowledge and practitioners when describing Mode 2. This is intended merely to highlight differences, not to suggest that practitioners of Mode 2 are not behaving according to the norms of scientific method. It is our contention that there is sufficient empirical evidence to indicate that a distinct set of cognitive and social practices is beginning to emerge and they are different from those that govern Mode 1. The only question may be whether they are sufficiently different to require a new label or whether they can be regarded simply as developments that can be accommodated within existing practices. The final answer to this question depends partly on acquiring more data and partly on how Mode 1 adapts to changing conditions in the economic and political environment.

Some attributes of knowledge production in Mode 2

Mode 2 knowledge production can be described in terms of a number of attributes, which can be used analytically to allow the principal differences between Mode 1 and Mode 2 to be specified. In Mode 2, knowledge is produced in the context of application. It is transdisciplinary, involves a variety of different skills in problem-solving and utilizes more flexible organizational structures. Mode 2 knowledge production is more socially accountable and makes use of a wider range or expertise in its quality control processes. Let us look at each of these attributes in turn.

Knowledge produced in the context of application

The relevant contrast here is between problem-solving which is carried out following the codes of practice relevant to a particular discipline and problem-solving which is organized around a particular application. In the former, the context is defined in relation to the cognitive and social norms that govern basic research or academic science. Latterly, this has tended to imply knowledge production carried out in the absence of some practical goal. In Mode 2, by contrast, knowledge results from a broader range of considerations. Such knowledge is intended to be useful to someone whether in industry or government, or society more generally and this imperative is present from the beginning. Knowledge thus produced is always produced under an aspect of continuous negotiation; this is, it will not be produced unless and until the interests of the various actors are included. Such is the context of application.

Application, in this sense is not product development carried out for industry and the processes or markets that operate to determine what knowledge is produced are much broader that is normally implied when one speaks about taking ideas to the market place. None the less, knowledge production in Mode 2 is the outcome of a process in which supply and demand factors can be said to operate, but the sources of supply are increasingly diverse, as are the demands for differentiated forms of specialist knowledge. Such processes or markets specify what we mean by the context of application. Because they include much more than commercial considerations, it might be said that in Mode 2 science is both in the market but also gone beyond it! Knowledge production, thus, becomes diffused throughout society, being produced in many different sites.

Research carried out in the context of application might be said to characterize a number of disciplines in the applied sciences and engineering – e.g. chemical engineering, aeronautical engineering or, more recently, computer science. Historically these sciences became established in universities but, strictly speaking, they cannot be called applied sciences, because it was precisely the lack of the relevant science that called them into being. They were genuinely new forms of knowledge though not necessarily of knowledge production because, they too, soon became the sites of disciplinary-based knowledge production in the style of Mode 1. These applied disciplines share with Mode 2 some aspects of the attribute of knowledge produced in the context of application. But, in Mode 2 the context is more complex. It is shaped by a more diverse set of intellectual and social demands than was the case in many applied sciences while it may give rise to genuine basic research.

The development of hypersonic aircraft provides an interesting if somewhat complex example of the processes that can take place in the context of application. The difficulties of taking forward the construction of a hypersonic aircraft, perhaps makes it as clear as necessary how inadequate it is to regard the context of application as the regime of applied science, because this science simply is not there. What is more, it cannot get started until certain technological discoveries are made that will structure the scientific domain so that appropriate equations will be not solved but *derived*. But the structuring of that domain cannot proceed without a whole series of previous judgments as to the shape of a desirable commercial vehicle. It is a good example of the interdependence of theory and design as well as commercial considerations, which in their interaction will produce pathbreaking science.

Thus, Foray and Gibbons, in analyzing the context of application have concluded that:

> [t]he hypersonic aircraft of the future will of course be a new artefact. This new artefact will not only embody much knowledge scientific and technical but it will also be a new design configuration of that knowledge. Both aspects are important. With regard to knowledge, the existing science and technology do not provide a sufficient base to guide the transition from supersonic to hypersonic velocities.. ... As far as hypersonics are concerned the primary need is for new infra-technologies and instrumentation to allow exploration of the new domain. There is a sense in which these must be available before the research agenda proper can be generated, but the research agenda even at this early stage cannot be constructed independent of the choice of design configuration and that involves interaction with a further range of engineering and commercial expertise. The ultimate mission, after all, is to produce a viable aircraft, not merely to elaborate a

theory or to produce a schematic design possibility. In brief, 'discovery' in this 'context' cannot be separated from the particular application that is reflected in the chosen design configuration which is itself partly a commercial matter. (Foray and Gibbons, 1996)

Transdisciplinarity

Mode 2 does more than assemble a diverse range of specialists to work in teams on problems in a complex applications oriented environment. To qualify as a specific form of knowledge production it is essential that inquiry be guided by specifiable consensus as to appropriate cognitive and social practice. In Mode 2, the consensus is conditioned by the context of application and evolves with it. The determinants of a potential solution involve the integration of different skills in a framework of action but the consensus may be only temporary depending on how well it conforms to the requirements set by the specific context of application. In Mode 2 the shape of the final solution will normally be beyond that of any single contributing discipline. It will be transdisciplinary.

Transdisciplinarity has four distinct features. First, it develops a distance, but evolving, framework to guide problem-solving efforts. This is generated and sustained in the context of application and not developed first and then applied to that context later by a different group of practitioners. The solution does not arise solely, or evenly mainly, from the application of knowledge that already exists. Although elements of existing knowledge must have entered into it, genuine creativity is involved and the theoretical consensus, once attained, cannot easily be reduced to disciplinary parts.

Second, because the solution comprises both empirical and theoretical components it is undeniably a contribution to knowledge, though not necessarily disciplinary knowledge. Though it has emerged from a particular context of application, transdisciplinary knowledge develops its own distinct theoretical structures, research methods, and modes of practice, though they may not be located on the prevalent disciplinary map. The effort is cumulative, though the direction of accumulation may travel in a number of different directions after a major problem has been solved.

Third, unlike Mode 1 where results are communicated through institutional channels, the results are communicated to those who have participated as they participate and so, in a sense, the diffusion of the results is initially accomplished in the process of their production. Subsequent diffusion occurs primarily as the original practitioners move to new problem contexts rather than through reporting results in professional journals or at conferences. Communication links are maintained partly through formal and partly through informal channels.

Fourth, transdisciplinarity is dynamic. It is problem-solving capability on the move. A particular solution can become the cognitive site from which further advances can be made, but where this knowledge will be used next and how it will develop are as difficult to predict as are the possible applications that might arise from discipline-based research. Mode 2 is marked especially but not exclusively by the ever closer interaction of knowledge production with a succession of problem contexts. Even though problem contexts are transient, and problem-solvers highly mobile, communication networks tend to persist and the knowledge contained in them is available to enter into further configurations.

Good examples of transdisciplinarity in the context of application can be seen in the development of new computer architecture. In these cases a diverse set of skills is required, including solid state physics, software engineering, linguistics, psychology, physics, philosophy and, of course, computer scientists. They aim to produce knowledge, which will be of use. These experts come together in what has been referred to as a context of application because they are challenged by the problem of the next generation of computer architecture and because they know that whatever the solution found, the next but one generation will begin from it. In a sense they cannot afford to be left out of the conversation. Further, such architectures are not the result of developing theory first and applying it later. Rather, theory must in some sense be as much the outcome of a successfully working computer as an input to its design. Some maintain, in the case of novel computer architectures, the theory, the design configuration and the relevant software are so intertwined that they emerge at the same time, rather like a new aeroplane which rolls out of a hangar accompanied by a flight simulator precise enough to train the pilots. It is simply too complex to design the aeroplane first and then build a simulator based primarily upon flight test information.

Heterogeneity and organizational diversity

Mode 2 knowledge production is heterogeneous in terms of the skills and experience people bring to it. The composition of a problem-solving team changes over time as requirements evolve. This is not planned or coordinated by any central body. As with Mode 1, challenging problems emerge, if not randomly, then in a way which makes their anticipation very difficult. Accordingly, it is marked by:

(a) An increase in the number of potential sites where knowledge can be created; no longer only universities and colleges, but non-university institutes, research centers, government agencies, industrial laboratories, think-tanks, consultancies, in their interaction.
(b) The linking of sites together in a variety of ways – electronically, organization-ally, socially, informally – through functioning networks of communication.
(c) The simultaneous differentiation, at these sites, of fields and areas of study into finer and finer specialities. The recombination and reconfiguration of these sub-fields form the bases for new forms of useful knowledge. Over time, knowledge production moves increasingly away from traditional disciplinary activity into new societal contexts.

In Mode 2, flexibility and response time are the crucial factors and because of this the types of organization used to tackle these problems may vary greatly. New forms of organization have emerged to accommodate the changing and transitory nature of the problems Mode 2 addresses. Characteristically, in Mode 2 research groups are less firmly institutionalized; people come together in temporary work teams and networks which dissolve when a problem is solved or redefined. Members may then reassemble in different groups involving different people, often in different loci, around different problems. The experience gathered in this process creates a competence which becomes highly valued and which is transferred to new contexts. Though problems may be transient and groups short-lived, the organization and communication pattern persists as a matrix from which further groups and networks, dedicated to different problems, will be formed. Mode 2 knowledge is thus created in a great variety of organizations and institutions, including multinational firms, network firms, small hi-tech firms based on a particular technology, government institutions, research universities, laboratories and institutes as well as national and international research programs. In such environments the patterns of funding exhibit a similar diversity, being assembled from a variety of organizations with a diverse range of requirements and expectations which, in turn, enter into the context of application.

Some of these aspects are illustrated by the Human Genome Project and that part of it which is concerned with the mapping of human genes. 'The aim of the Human Genome Project, quite simply, is to draw up a catalogue of our entire genetic make-up – the genome. The maps, like geographical maps, can be of varying type and resolution, from large-scale linkage maps of genes in relation to other genes based on frequency of co-inheritance, through various types of physical maps that locate "landmarks" in the DNA, and eventually to the highest resolution, the sequence of chemical base-pairs which make up the DNA molecule. Proponents of the project claim that it will provide a valuable resource for science and medicine, while opponents have challenged its wisdom in terms of cost, strategy, ethics and the ultimate utility of its results' (Balmer, 1996).

But the fact that a mapping project emerged can only be understood as the outcome of a complex process of negotiation in which a large number of interested parties were involved. Some actors and institutions played a large part while others were less important. The point is that no one single person or groups were in control dictating the pace and direction of advance. For example, Balmer concludes that, 'the HGMP was not caused by any single factor. The state policies of selectivity and concentration, and value for money were co-ordinated with the agendas of the MRC and the gene mapping community and their spokespeople. There was no need for different groups to identify with each other's agendas. The project acted as a boundary object in this respect, a political entity that aligned separate working goals and agendas. Alignment was achieved over a period of time as groups and their agendas were shuffled into and out of the policy arena, or altogether marginalized. As a consequence, money flowed from the state to scientists, and gene mapping under the auspices of a concerted organised programme came to be supported' (Balmer, 1996).

'At each stage of the development of the project, however, the contingency of events and the opportunism of the actors cannot be ignored. Brenner and Bodmer may have had some degree of control; the government, ACOST civil servants, and MRC administrators may have had a degree of control. But in sum it was more of an orchestration process, and making the best use of the resources available than a planned strategic, networking process that led to the HGMP in the United Kingdom. In this case, it was rather like having an orchestra where all the players are vying to be the conductor, with no one fully in control and every one ready to improvise' (Balmer, 1996).

In the case of the Human Genome project one can observe the role of the context of application under which the project overall was carried forward, the complex set of negotiations which took place over an extended period of time, as well as a novel form of organization – a boundary object – which acted as a vehicle around which many diverse interests could converge, if not completely, at least enough to allow funding to flow.

Social accountability and reflexivity

In recent years, growing public concern about issues to do with the environment, health, communications, privacy and procreation, and so forth, have had the effect of stimulating the growth of knowledge production in Mode 2. Growing awareness about the variety of ways in which advances in science and technology can affect the public interest has increased the numbers of groups who wish to influence the outcome of the research process. This is reflected in the varied composition of the research teams. Social scientists work alongside natural scientists, engineers, lawyers and businessmen because the nature of the problems requires it. Social accountability permeates the whole knowledge production process. It is reflected not only in interpretation, and diffusion of results but in the definition of the problem and the setting of research priorities, as well. An expanding number of interest groups, and so called concerned groups, are demanding representation in the setting of the policy agenda as well as in the subsequent decision-making process. In Mode 2 sensitivity to the impact of the research is built in from the start. It forms part of the context of application.

Contrary to what one might expect, working in the context of application increases the sensitivity of scientists and technologists to the broader implications of what they are doing. Operating in Mode 2 makes all participants more reflexive. This is because the issues, which forward the development of Mode 2 research, cannot be specified in scientific and technical terms alone. The research toward the resolution of these types of problems has to incorporate options for the implementation of the solutions and these are bound to touch the values and preferences of different individuals and groups which have been seen as traditionally outside of the scientific and technological system. They can now become active agents in the definition and solution of problems as well as in the evaluation of performance. This is expressed partly in terms of the need for greater social accountability, but it also means that the individuals themselves cannot function effectively without reflecting – trying to operate from the standpoint of – all the actors involved. The deepening of understanding that this brings, in turn, has an effect on what is considered worthwhile doing and, hence, on the structure of the research itself. Reflection of the values implied in human aspirations and projects has been a traditional concern of the humanities. As reflexivity within the research process spreads, the humanities too are experiencing an increase in demand for the sorts of knowledge they have to offer.

Some effects of reflexivity and enhanced social accountability can be seen in the outcomes of various fora for the expression of public concern. Of particular interest is the role played by public controversies in the generation of markets for novel knowledge and expertise. Public controversies create meeting places for discussion. Because many diverse actors are involved these meeting places can be regarded as hybrid fora. Controversies frequently lead to establishment of inquiries dealing with questions of public policy, regulation and a host of other social and ethical issues.

New knowledge is gathered, some of it based on the results of previous scientific research and technological developments that have gradually become the cause of social concern. For example, new forms of knowledge such as risk analysis, technology assessment or the growth of various specialisms in environmental science, are responses to public concern about the safety of high risk buildings, the adverse effects of automobile traffic or the effects of global warming. Through these controversies markets for alternate technologies are developed and foci for new research agendas established. This role of public enquiries in generating new research in environmental science has been elaborated by Cambrosio and his colleagues (Cambrosio *et al.*, 1992).

Quality control

Criteria to assess the quality of the work and the teams which carry out research in Mode 2 differ from those of more traditional, disciplinary science. Quality in Mode 1 is determined essentially through the peer review judgments about the contributions made by individuals. Control is maintained by careful selection of those judged competent to act as peers, which is in part determined by their previous contributions to their discipline. So, the peer review process is one in which quality and control mutually re-enforce one another. It has both cognitive and social dimensions, in that there is professional control over what problems and techniques are deemed important to work on as well as who is qualified to pursue their solution. In disciplinary science, peer review operates to channel individuals to work on problems judged to be central to the advance of the discipline. These problems are defined largely in terms of criteria, which reflect the intellectual interests and preoccupations of the discipline and its gatekeepers.

In Mode 2 additional criteria are added through the context of application which now incorporates a diverse range of intellectual interests as well as other social, economic or political ones. To the criterion of intellectual interest and its interaction, further questions are posed, 'Will the solution, if found, to competitive in the market? Will it be cost effective? Will it be socially acceptable?' Quality is determined by a wider set of criteria that reflects the broadening social composition of the review system. This implies that 'good science' is more difficult to determine. Since it is no longer limited strictly to the judgments of disciplinary peers, the fear is that control will be weaker and result in lower quality work. Although the quality control process in Mode 2 is more broadly based, it does not follow that because a wider range of expertise is brought to bear on a problem that it will necessarily be of lower quality. It is of a more composite, multidimensional kind.

These attributes set imperatives for institutional change and pose serious challenges to the conventional organization of research wherever it is conducted. First, it is important to grasp that it is not being argued that the new practices are going to eliminate the old; that Mode 1 will eventually succumb to Mode 2. It is far more likely that both will continue to co-exist. The terms of that co-existence depend as much on the response of the institutions that are currently supporting Mode 1 as on the social diffusion of Mode 2. The point is that further articulation of the socially distributed knowledge production system cannot but challenge most established knowledge producing institutions. In particular, if institutions do not adapt they will be bypassed but all the signs are that some of them, particularly the better universities, are adapting.

A socially distributed knowledge production system

The transformation of knowledge production, in the sense described above, is one of the central processes characterizing the societies of the advanced industrial world. Knowledge production is less and less a self-contained activity. It is neither the science of the universities nor the technology of industry, to use an older classification for illustrative purposes. Knowledge production, not only in its theories and models but also in its methods and techniques, has spread from academia into all those institutions that seek social legitimation through recognizable competence and beyond. Science is less the preserve of a special type of institution, from which it is expected to spill over or spin-off to the benefit of other sectors. Knowledge production is increasingly a socially distributed process. Moreover its locus is global, or soon will be. At its base lies the expansion of the numbers of sites which form the sources for a continual combination and recombination of knowledge resources; the 'multiplication of the nerve endings of knowledge' that is discussed repeatedly in this book. The expansion of the number of sites where recognizably competent research can be undertaken has implications for the management of the knowledge production process and for the maintenance of quality control within it.

The distributed character of knowledge production constitutes a fundamental change. To it are linked the other dimensions of change which we have explored: the increasing contextualization including the marketability of knowledge, the blurred boundaries between disciplines and institutions and across institutional boundaries, fungibility of scientific careers, transdisciplinarity not only of hot topics, increasing importance of hybrid fora – groups constituted through the interplay of experts and non-experts as social actors – in the shaping of knowledge.

The continuing massification of the university as a teaching institution is a prerequisite for this wider distribution in society of the capability to produce and use knowledge. But, as already argued, some changes in emphasis are necessary. In particular, students need to learn how to find, appropriate and use knowledge that might have been produced almost anywhere in the world. However, it is in the adaptation of their research function to the distributed character of knowledge production that universities are most challenged. The university must enlarge its view of its role in knowledge production from that of being a monopoly supplier to becoming a partner in both national and international contexts. Such a change will, before long involve a redefinition of excellence amongst academics, of their career aspirations, of their disciplinary contributions, and their institutional loyalties. The universities, in their turn, will need to explore strategies of niche specialization.

In industry, too, the distribution of knowledge producing capability raises profound questions about the appropriation of that knowledge for its own purposes; that is, about the extent and organization of industry's in-house R&D as well as about the ownership and management of intellectual property and acquisition of the skills needed to configure knowledge resources in a manner relevant to industry's competitive situation. Resourcefulness in the management of the configuration of knowledge is a precondition to successfully playing for success in a world where the intensification of competition in markets internationally, together with the transformation of the information infrastructure have made technological innovation the name of the game.

This analysis of the transformation of knowledge production indeed entails major changes in approach to policy. Although this volume has been concerned with knowledge production in its broadest sense, including the humanities, the changes in policy orientation that are now required can be seen more sharply by focussing on scientific and technological knowledge. Therefore, we will explore the significance of the shift in the mode of knowledge production against the background of the development of science and technology policy over the past half-century.

As has been indicated, many national science policies continue to be dominated by what has been called 'linear thinking'. In the models which emerge from this thinking, science functions as the source of technology and the engine of economic growth. In the linear model, the universities and some government research laboratories are paramount, being the institutions which carry out most of the basic research. With the support, research function of these institutions has been developed. It is the model that currently dominates the institutional horizon and the one with which we are most familiar. National higher education policies, too, reflect the same linear thinking. Universities are expected to contribute a share of the ideas, on which national competitive advantage will rest and train the requisite scientific and technical manpower necessary to operate a modern economy.

The emergence of Mode 2 puts the validity of the linear model further in doubt. In so doing, it calls into question not only the presupposition of national science policies but also that bit of higher education policy which touches on research and the training of qualified manpower. The university, in the emerging regime, must still be an instrument for the development of science. The point is that it is no longer either the only or even the primary institution on the cognitive landscape. The emergence of a socially distributed knowledge production system brings to the fore the question of the relationships between the university and the other knowledge producers. If knowledge production at the leading edge has the attributes of Mode 2, then universities will need to become more porous institutions, more revolving doors are required which allow academics out and others in. Such development if carried out on a significant level cannot but touch questions of career development and reward structures, and with this challenge the existing structures.

Governments and the management of distributed knowledge production

What does a policy for distributed knowledge production look like? At the very least, it requires some radical departures from the traditional viewpoint. First, the notions of separate science and technology 'markets' have to be abandoned since actors do not move in accordance with linear, sequential, and hierarchical models, step by step from research, to development, to innovation and use. Basic science has become inseparable from technological development linked by the innovative use of instrumentation. It has been conventional to view the frontier of science as expanding from the core of its activities. In the current context, however, both core and frontier are spreading. This is evident in such areas as molecular biology, biotechnology, new materials, nano-technology, liquid crystal and solid state physics, nuclear fusion, informatics, superconductivity.

Second, the new policy models are no longer of the 'systems' type popular a decade ago among policy analysts. Systems models imply greater stability in the relationships between actors than is justified given what we already know of distributed knowledge production. More helpful descriptions could be worked out by trying to develop models that incorporate the evolution of patterns of interconnections, the ability to establish, on a recurrent basis, new modes of exchange, the skills to adapt to the richness of research practice, and to create ever new channels of communication.

Third, specialization takes on entirely new forms. They are not to be understood as a further division of labor inside already constituted disciplines. The new specialisms which drive discovery and innovation are problem-oriented and mostly transdisciplinary in character. They break with the common vision of specialization as an incipient discipline or sub-discipline that is starting on its way to professionalization and institutionalization. They exhibit much more mobility. They are tied to the resolution of clusters of problems and will develop in accordance with new problems.

Fourthly, the policy process will need to develop a new management style. The traditional approach – some variant on management by objectives like the systems approach – is too inflexible. The management of a distributed knowledge production process needs to be open-ended, and to break away from classical planning perspectives. The management of processes, particularly of the external environment becomes paramount. That management style can be summarized in two notions – increasing permeability of boundaries and brokering.

Permeability

In distributed knowledge production the dynamics of science and technological innovation are the principal driving forces leading to the emergence of new forms of organization. While some universities and research institutions have been slow to adapt, the best have already become more permeable and integrated into new networking arrangements. The process of increasing permeability of boundaries weakens the centralizing tendency of bureaucracy. Policies of decentralization should incorporate incentives to encourage openness and reward individuals who can achieve economies of scope with existing resources. Large university-type institutes with tenured facility, or government laboratories for fulfilling specific functions as well as permanent research units with tenured research staff set up for specific monocultural research will not provide the policy models of the future. Such organizations have become too expensive and inflexible to meet the needs of distributed knowledge production.

Alternative models might involve the creation of lean 'centers', employing few administrators with a budget to stimulate networks of innovators, in units attached to diverse institutions, agencies or firms. They would be periodically evaluated in terms of their effectiveness in process management. When their jobs were completed, or when decreasing returns became evident they would be disbanded. These centers, like other institutions, created in the context of socially distributed knowledge production are likely to have many stakeholders and will need to be run and evaluated accordingly. Any policy that tended to entrench institutions, or encourage autarkic attitudes, is anachronistic.

It is well known how to set up laboratory sites to pursue scientific investigations of various kinds. We also know how to build teams around single professors and individuals of exceptional talent, e.g. Max Planck Gesellschaft in Germany. What we do not know so well is how to manage the art of facilitating efficient communication between such nuclei as well as between the other equally important elements that one finds in Mode 2. This is a matter not only of facilitating relationships between groups of researchers but also of making easier communications with innovators, regulators, venture capitalists, etc.

Brokering

The second notion in this phase of science policy should be for government, alone or in cooperation with others, or some of their agencies, to function as honest brokers. Governments are a logical choice for this role because much of the brokering is likely to involve other governments or their agencies. Brokering is necessary because in distributed knowledge production more actors, not all of whom are technical experts, are involved. Brokering will demand exceptional skills because the individuals involved in the innovation process will come from many different institutions and organizations, they will often be dispersed geographically and may only be able to work on a problem or project part-time. Distributed knowledge production is diffusing rapidly because congenial settings embodying a large variety of organizational styles are being established. The task of policy is to provide the framework for the management of this flux.

This implies that the policy arena itself will undergo a drastic change in composition. Indeed, this has already begun. During the phase characterized as policy for science, it was expected that academic scientists would be key policy players. This has become much less so, as governments have shifted from the support of science for its own sake toward innovation policy. During the last two decades not only politicians and civil servants, but also economists, marketing experts, and industrialists, have become involved in the genesis of science and technology policy. This intrusion of the wider interests of society is sometimes resented by scientists because it is felt to erode the independence of the 'Republic of Science'. But there are good reasons for the shift in the locus of authority in the development of science: it reflects the distributed nature of knowledge production. Adapting to this change will be a major challenge for universities and academies alike.

Issues related to risks for health or the environment posed by technological development, or issued related to the impact of information technology on jobs, training and competence, or ethical issues brought out by new biomedical technologies, are becoming matters for public debate. In short, the new innovation policy is now, inescapably, a part of politics.

However, attitudes and processes characteristic of earlier phases of science policy will survive. We are describing, not a historical break, but rather a significant shift of emphasis. Some priorities will continue to be set within the scientific establishment –

resource limitations will make this imperative. Also efforts will continue to be made within the scientific industrial system to filter science through the sieve of industrial needs. Arguments about priorities and about protecting the national science base will continue but the agenda is no longer being set primarily in universities or in national research councils. Though scientists remain the driving force in proposing areas for research, the research priorities will be generated within hybrid fora composed of many different actors. These priorities now have both a cognitive and a social dimension. Innovation policy will be concerned not with the details of this process but with the direction and support of the multiplication of knowledge production sites and the management of complexity that results from acceleration of knowledge interconnections.

This new style of management is not required simply because of the increase in the numbers of new sites of knowledge production. It is also necessary because of the transformed nature of the aims and content of knowledge production. The two processes are interconnected. The changing nature of what is regarded as important in science and the new requirements of its production have induced the organizational changes described earlier. The close interaction between form and content in knowledge production is the primary reason why the new tasks have become those of handling and processing knowledge. This requires new approaches to creativity, strategies for preserving and increasing the permeability of institutions, for designing and intensifying inter-linkages among them, for managing flux rather than administering institutions old and new.

But all this rests ultimately on *a policy that is people- and competence-centered.* This new innovation policy will need different institutions, although some existing institutions will adapt more successfully than others to a new style of management. It will also require a new competence to create and design, rather than to reproduce with marginal improvements. This is one of the reasons why we stress the importance of developing policies that promote interchange amongst scientists and technologists and the general connectivity of innovation systems, possibly using information technology to exploit its knowledge base. The competence – the new skills and perspectives that emerge from these interchanges – is at least as important an outcome of this mode of knowledge production as the problems solved or the artefacts created. This implies policies that promote transdisciplinarity and provide for the possibility that unusual modes of organization may be required; policies that promote international collaboration and that seek to be aware of, and to be able to interpret, knowledge wherever it may be being produced. In managing this flux people as the carriers of competence will constitute the main resource.

The new policy will also have to manage better the interface between competition and collaboration. Managers of innovation policy will need to shift continuously back and forth between two different frameworks, of competition and collaboration. The job of governments, of research managers and others is not to pick winners, whether individual products or generic technologies, nor simply to create an environment which encourages rivalrous behavior. So far, international competition occupies center stage in most national innovation policies and will continue to so do for the foreseeable future. Competition generates diversity by provoking rivalrous behavior amongst the competitors. In the production of knowledge this implies the ability to experiment. Dynamic competition is essentially a discovery process, it leads to innovation. But innovation also reduces diversity, and rivalry, at certain stages of the process, can become dysfunctional. It is the function of collaboration to restore diversity. Rivalry itself is not enough. In Mode 2 knowledge production, unrestrained competition can have the effect of inhibiting the growth of networks and of discouraging permeability. Managing the delicate transition between environments appropriate for competition and those appropriate to collaboration implies very complex boundary conditions involving different types of institutions that may be globally distributed. Competition policy has to become competition and collaboration policy, dynamically conceived; not in terms of setting the stage for a zero-sum game, but as creating the environment in which diverse forms of behavior and organization can flourish.

The policy has implications on the sensitive problem of international competition and international strategies of collaboration. Far-sighted governments will best pursue their own national interests, not by seeking to develop self-contained national science policies, but by encouraging the growth of international networks and permeability. Success will be determined largely by the ability to extract economic value out of international collaboration. It need not be a predatory act because it is not a zero-sum game.

The new policy will also have to confront a potential imbalance between the volatility and permanence of institutions. It may seem that much of what this chapter has said until now will generate only micro-instability. But self-organizing systems also produce eigenvalues; that is regions of macro-stability. Rather than preserving short-term micro-stability by supporting specific firms or sectors, innovation policy needs to aim at long-term sustainable economic development. This is achieved by encouraging diversity, promoting experimentation and creativity, facilitating the emergence of open environments for interaction and exchange. Since this policy will be built upon the importance of people and competence in the innovation process, the institutions in which people work will have to find a balance between permanence and change. Institutional managers need to shift the balance from existing, stable and continuous forms of organization to those which are more flexible and temporary. But it is a question of finding a middle ground between rigidity and chaos.

CHAPTER FOUR

Government Policy for Industrial Innovation

Lewis Branscomb

Introduction

When we think about policy for research and innovation, we should be less concerned with the different 'kinds' of research that usually fall under that rubric. In the US we are very caught-up on the debate between economic conservatives and liberals, over whether or not the government has any business in funding R&D programs in the commercial/private sector, never mind the military. So we, in America, struggle and the most controversial of those programs is the Advanced Technology Program (ATP), which the 104th Congress newly validated but which the Republicans tried to abolish. Obviously in every country there are big issues about public investments in both research and human resources. This, afterall, is core to what the government's role is. But there is also a very important government role in encouraging private sector innovation and the tools for doing that are quite different than those used for subsidizing research as Michael Gibbons and John de la Mothe have each suggested in previous chapters. Along these lines, I will specialize my remarks in this chapter to deal with economic goals for research and innovation policy. We do not have a problem with defining what the government's goals are with respect to economics: providing a certain quality of life, as well as security and opportunity for people; capturing benefits domestically in a global economy; and so on. A very important element of policy is to resolve the question of choosing to be better than the other guy or choosing to be in an absolutely better position while not dominating internationally. A great many aspects of US economic policy certainly depend on varying political attitudes toward whether the economic pie is viewed as fixed or not.

Policy change and challenge

It is important to recognize that governments change policy rather slowly, particularly on something as arcane as research and innovation. But the world does not stand still. Competitiveness clearly depends more now on research-based innovation. In nearly all of the most advanced economies, firms spend more of the aggregate on R&D than on capital investment. Foreign firms in many countries, including some emerging economies, are very adept at absorbing the latest technologies, this being because so much of engineering and science has become codified – that is, based on a theory which allows one to quantitatively model and simulate technology. In other words, technology flows across national borders very rapidly. Large national firms are now no longer giving their suppliers the drawing to which they are to make the components, but are giving them functional specifications and are challenging them to innovate, thereby unleashing the enormous innovative capacity of small to medium sized industry. Those companies which tend to be technologically defined, selling their components in many markets, always look to global markets; since they have defined a technology expertise, they are really interested in the maximum number of areas into which they can put their niche expertise. This is a formula, which in fact is driving a lot of very high level innovation. So I think the future lies very strongly in every hub of economic activity with a relatively small number of big multinationals who are very good at accessing markets, production, distribution and marketing channels, and who then outsource component innovation and then integrate the systems into the ultimate products. Little firms have the world as their oyster, not just the country they live in.

Finally, foreign direct investment plays an important role in the technical capacity of every country, and it is interesting to note that in the US foreign direct investment in R&D establishments is the most rapidly growing segment of R&D growth in the US and has been so for some years.

So, what is the government's policy problem? Part of the problem is sorting out the logic for federal funding and research. Why does the government buy research activities? Is it to improve the economy or is it to improve the culture? After all science is in some part an expression of the best of our culture. What kind of research should the government fund? Should it fund pure science, should it fund technology development, engineering, etc.? And who should do the work? How do we divide up between universities, government laboratories, government-sponsored laboratories and industry? And if we knew the answer to these questions, who pays for what? Government, industry, R&D, or foreign firms? These are the core questions to be answered, in the governance of science and technology.

On the innovation policy issue, what should the government do besides subsidize R&D? Because scientists, in the US at least, are the political constituency that tries to improve our R&D budgets, we get the impression from the political debates (at least from those who are down in Washington trying to get the attention of Congress) that the only thing that really matters is what the government spends on research, but government spending on research is absolutely crucial to the future capacity of any nation. Government also wants people to enjoy the fruits of that research. So every time the government sets trade policy, economic policy, intellectual property law, environmental standards and regulations, these issues ought to be looked at from the point of view of how do they impact on innovation. Many of them impact innovation very negatively because those who set these policies are not charged with thinking about their impact of innovation. So I prefer to speak about research and innovation rather than science and technology in order to make sure that if we do the research we train the people, and that they have the creative capacity so that the whole social system can actually deliver the benefits to the public. And this can be delivered only through good innovation policies. Thus, the question of what should government pay for? What government interventions in the research marketplace are useful economically yet will not distort functioning markets? This is important in any capitalist society that really believes in the power of markets to allocate capital and other resources efficiently, and suggests that the government should invest when there are market failures.

Market failures

What are the market failures? I would suggest that there are two kinds. First, there is the one best described by micro-economists as the tendency to underinvest in R&D in relation to the public values that come from private R&D investment. Economists like Ed Mansfield have done a lot of research to show that the public return from private investment in R&D is somewhere between 20 to 40 percent and the public return from public R&D is even higher.

Secondly, and maybe even more important, is a whole set of issues involving failures in social capital. The issues of if you do the research very ably, how does that knowledge diffuse throughout the society, how do institutions and individuals adopt it, or absorb the new ideas and derive benefit from it are of vital importance. If one looks at the Canadian ratio of government investment in diffusion and absorption compared to creation, it will be found that Canada is much stronger relative to the US. I think this is due to a failure of US policy, not due to Canada's policy.

These are the two sets of market failure. With respect to the economic causes of underinvestment, there is certainly an opportunity cost associated with taking a technological risk; if the risk turns out badly, time and resources have been spent on something that has not resulted in a product. One has given up the opportunity to have done something less risky, like an incremental improvement of an existing product line, that would have at least brought in some revenue. When thinking about some new technical idea that might be developed into a product, most entrepreneurs have trouble defining and sizing the market before developing the actual technology. This inability to define the market with confidence really does reduce the conviction that the technology risk is worth taking. As mentioned above, the private returns may be significantly less than the public returns. And then there is the very interesting issue or problem of picking winners – this is the skew in returns that makes the average much less attractive than the best returns, so it really matters what you choose.

I have begun to focus on a second category of market failure that justified government investment revolving around the question of how entrepreneurial individuals in high technology companies think about technical risk and how they deal with it. One of the important issues is that social psychological effects may deter firms from taking sufficient technical risk. Furthermore, these risks are quite different for different sizes of firms.

Firm size

Small firms do not have the debt-to-resources ratio necessary in order to take a bunch of technical 'flyers' that fail to pan out. They really have to guess right because they have one shot at the mark. On the one hand they want to be very imaginative and aggressive because they are a small start-up but on the other they are forced to be conservative owing to the nature of that situation. This is compounded by the fact that venture capital companies tend to follow fads, which is entirely irrational. Five years ago a small biotech start-up would have been unlikely to have had any problem getting money. Today there are biotech start-up companies with really fabulous ideas and with really talented people that cannot get venture money because the venture money is going to the software companies. There is no justifiable reason for this behavior.

Medium size firms have the problem that venture capitalists insist on knowing what the market is for the business and they want to see the technology already reduced to its practical use before they invest. So they want to see the really tough stuff already done before they put money in.

Finally, large firms tend to be risk averse. The problem is that if within the company there comes an entrepreneurial idea which if successful may have 100 percent growth per year over the next ten years, the first year's market may be $1 million so it will be ten years before it is $100 million. And if this is a $50 billion company it will not even be the width of the black line on the income chart and so the business planners will say why in the world should we put money into this risky thing when it will not generate enough revenue to make a real difference. This is a serious problem.

Social capital

There is also the crucial problem of the ability to capture and absorb the outcomes of innovation flows. A recently completed project on Japanese/American university–industry economic relationships (Branscomb *et al.*, 1999) showed very clearly that it is enormously important that the social capital is in the system to allow universities and high-tech companies with inventions to create new opportunities and to capture the benefits locally. In the US this is a regional issue. Cleveland, Ohio, for example, is a wealthy and productive industrial area in which companies spend $2 billion per year on R&D but it's a 'rust belt' economy none the less, which does not have a great future. So Cleveland is working on the problem of how to switch over to a high-tech, high value added economy. (Cimioli and de la Mothe discuss the economics of this in a later chapter.) There is a committee of top chief executives representing the companies in Cleveland that advises the mayor and the governor, and there are special Ohio programs to assist the universities in the Cleveland area to set up intermediate institutions to provide practical help so as to develop and assist start-up industries. Bottom line? They have made great inventions and have started great companies. The problem is that out in California, up in Boston, in Japan and in Canada there is a bunch of really smart folks ready to run when they see a new idea and that Cleveland idea gets snatched up and run with before Cleveland can figure out what happened. The problem is social health. On the other hand we have a study in Massachusetts by Professor Adam Jaffe that shows that every dollar in federal R&D brings in immediate private sector return in large multiples and every decrease in federal R&D will bring a decrease. This is because in Massachusetts, thanks to MIT and some other institutions, there is a tremendous ability to absorb innovative ideas and capitalize on them.

Looking at technical risk in commercial investment, conceptually, R&D spending rises to a maximum around the third year of the program and after the fourth year, which is the point of the market roll out, one is still doing some R&D, but this spending falls off (decreases) after another (fifth) year. The level of uncertainty against time that goes with that investment shows that it is necessary to prove the technical feasibility within the first year. At that time work is already taking place on product performance, on product-cost risk, on market acceptance risk, etc. But the issue at this stage is to demonstrate that the technology is for real. (See Figure 4.1.)

The second issue that has to be demonstrated is that the technology does what it was hoped it would do. The third issue is to prove that it can be done within the planned or anticipated cost. The proof that cannot be obtained until one gets to market is, is the market real? If that is in fact the way entrepreneurialship works in high risk technical industry, then the issue for government is what can the government do to eliminate those sources of risk? It is completely inappropriate for the government to try to eliminate market risk – unless it is a case where the government makes the market, in which case that is the best way to eliminate risk – otherwise, the best chance for a government is to do something about technical risk. This leads to the Advanced Technology Program and the Technology Partnerships Canada Program, which is similar. In my view the important idea is a cost share partnership between the government and the private sector to try to reduce innovative ideas to practice. So the question is, 'can a government program help reduce the skew by helping firms to select the best ideas?'

'What are the effective mechanisms for diffusion?' – because if the benefits of this investment do not diffuse into the economy, then the government can be accused of subsidizing one company to compete with others and there will be a political reaction to that.

Finally, an interesting question is, 'is earnings recapture appropriate?' The ATP started off with that goal in mind but the law has been changed so that the program does not have to recapture. The basic argument on the US side was simply that if in fact a project is co-funded partly by government and partly by the company, and it was believed by all the parties that the public benefits were at least equal to the private benefits and the ratio of cost sharing somehow represented the ratio of benefits, then indeed it is unnecessary for the government to recapture. However, one can only make this argument on an aggregate statistical basis. There is no way that any one project could ever demonstrate whether that fifty-fifty cost sharing was the correct ratio, it probably should have been one-zero or zero-one for most outcomes. So earnings recapture really solves a political difficulty.

Breaking through the linear model

Then there is an interesting question about the rest of the government portfolio, other than cost sharing research with companies. In my view, in the US we have become trapped in a silly position. The Congress and even (maybe even especially) the conservatives agree that the government is the only institution that can fund basic science in the universities. Those people say 'well that is basic research – but if the research is demonstrably useful to someone, then whoever it is useful to surely will pay for it'. The end result of that syllogism is that one has members of the Congress saying that the government should fund useless research, but not any of that other stuff. This is clearly ridiculous.

How did we get ourselves in this position? We did so because we confused the motive for making an investment with the circumstances under which research should be done. My view is that basis research should not be defined by utility, but by the uncertainty as to who benefits. If one knows who benefits, then charge them for the work. If one does not know who benefits but can prove that it is the public that generically benefits, then there may be an opportunity for public investment. That leads me to believe that there is not only basic science research, but basic technology research as well because there is plenty of technology research that is cerebral whether in an academic setting or in a company. This is also deserving of government support, and the engineers should stop claiming that they are scientists because that is the only way they can get money. They should say that they are creative intellectuals who are doing something really important that returns benefit to the public and that they should be supported. My goal is that if I ever win this argument, instead of someone saying basic science research and basic technology research as separate words, we just say research. Government funds research and industry does development.

So who pays? That syllogism tells you that the intended beneficiary should pay and if that is basic research then the government has to pay. It if is research to procure for the federal government then the government should pay 100 percent plus overhead. If it is technical development that leads directly to commercial products, then firms pay 100 percent with no government funds. But when there are both private and public benefits then maybe cost sharing and collaboration is appropriate. So, the question is who does this publicly funded research? Obviously the most competent institutions, of which there are many. The second criteria is to ask what kind of institutional environment will give the most effective diffusion of the benefits throughout the economy, because that is what determines the ratio of private and public benefits. My view is that the universities have a uniquely good mechanism for diffusing benefits. The university's graduates carry the knowledge with them, which is instantaneously transferred to their employer. Our national labs have cooperative research and development agreements, which are sub-optimal effective and I think that this is an awkward system. The state governments have substantial regional development programs and the politics is very different from the federal level. For instance it is perfectly acceptable to fund an industrial economy and industrial development at this level. Industrial policy at the state level is political goodness not badness because the states compete with each other. The states do indeed invest for economic purposes, they have many programs to promote innovation, some of them are quite effective and my notion is that the federal government should link up with those state programs and leverage them as a means of not only bringing consortia of companies but also universities into the act. Firms, of course also diffuse technology internally and through shared research and industrial alliances.

This leads us to private-public partnership. These have a role in many places where the government is a customer, such as defense where we move to dual-use technologies and recognize that the defense department must leverage private technology which attracts much more money and has much more urgency than the government. The government funds basic technology research in its partnerships and emphasizes consortia to accelerate diffusion. The cost sharing should reflect the share of benefits and we should leverage the state investments.

So how much should the government invest? This is a much harder problem to solve. First of all let us agree that the motive of the investor, namely the government, should determine the investment and not the nature of the work. We can basically divide government R&D investment, except for putting aside the role of military procurement, into two categories: 'opportunity' driven research and 'needs' driven research. Mission agencies do needs driven research, as does the Department of Energy which funds research because they have an energy mission. The opportunity driven research is evaluated by peer review in horizontal competition with the researchers themselves deciding what is most promising.

Conclusion

In summary, first of all it is important to test all government economic, trade, intellectual property and regulatory policies for their effect on the innovative capacity of the private sector. It is necessary to recognize that basic technology research is a vital area of the research community that is needs driven. Those needs must be articulated by the private sector because they are not the government's needs. The government's narrow needs are not an adequate reflection of what is exciting in the commercial world; they seriously undervalue process research. They also seriously undervalue some parts of material research. So we need a mechanism for bringing technical judgment into those allocations. The failure to innovate is as often caused by management failure as by market failures and we need very flexible, industry specific policies because different segments of industry behave differently and have different structures, dynamics and sizes. Most importantly, intellectual property plays radically different roles in different parts of industry and this makes a big difference to what governments should do. We need to leverage global innovation and research. But, as stated at the outset, I do not believe there is such a thing as a 'national system of innovation'. Any system tends to be globally- or industry-specific. Every small high-tech, specialized company should view the world as its market. Finally, we all need to help government improve its ability to plan and manage this kind of policy and its needs, because we are talking about research and innovation which is a means to an end, not an end in itself. Voters care about health. They care about security. They care about the environment and they care about the economy. Science and innovation are critical means to meeting these concerns.

PART III

Evolving Debates in Science Policy and Public Administration

CHAPTER FIVE

Science, Technology and the Tools of the Politico-administrative Trade

Jane Marceau

Introduction

The preceding contributors have all written articulately about the science–government relationship. However, I will begin this part of the book by noting that governments almost everywhere are subject to increasing public scrutiny. The political consensus that was built after the end of World War II was built on the notion that governments could remedy the problems caused by unrestricted capitalism as seen in the Depression of the 1930s. This sense has large crumbled. Governments were then to be the agents both of economic growth and social justice, they were to stimulate and guide the development of the economy and to ensure that the fruits were distributed in a way which enabled the citizen body as a whole to gain maximum benefits. Governments were to use the taxation system to provide, on the one hand, public education, which is of benefit to both the economy and citizens, and, on the other, to put in place systems of protection for citizens against the risks of unemployment, ill health and old age.

Over the first decades after 1945 this system of 'mixed economy' worked well, providing the citizens of OECD nations with both prosperity and social and economic security. After the oil shocks of the 1970s, however, uncertainty increased about the path to development which involved such a central role for governments. While rates of unemployment rose dramatically throughout the 1970s, public authorities seemed unable to respond effectively.

The climate of uncertainty encouraged the re-emergence of voices claiming that the problems stemmed from 'too much interference' by governments with the 'mechanisms of the market', from the building of 'too strong' a welfare state which 'crowded out' capital and entrepreneurial energy needed to restore the health of the economy and reduce unemployment.

This sentiment was expressed through the ballot box, resulting in the election of right-wing governments in much of Europe and in the US by the end of the decade and during the 1980s. While there were some exceptions to this trend, as in France where a left-wing President was elected in 1981, and only small proportions of voters shifted their votes, the right-wing ascendancy was widespread. Even where labour-oriented parties were elected to power, as in Australia in 1983, the policies pursued owed much to the rhetoric and analysis, which advocated deregulation, privatization of public assets and a greater reliance on the 'market' in allocating resources. The welfare state came under threat everywhere. That threat has mostly increased in the 1990s.

In some countries, especially in the US, the form of opposition to the involvement of the state in regulating economic activity and the distribution of rewards has taken the form of a push for governments to 'do more with less'. Governments are to 're-invent' themselves so as to be more 'responsive' to the needs of citizens (Barzelay, 1992; Osborne and Gaebler, 1993; Sandel, 1996). Reduction of bureaucracy has become the cry of the era. States, citizens and individuals no longer always seem to have common interests.

Indeed, some commentators have redefined the whole notion of public interest. Barzelay, for example, recently wrote in a widely read volume that

> the specific rhetorical phrase 'the public interest' should be confined to books on the history of American politics and administration. A desirable substitute expression is 'results citizens value' ... [which] conjures up ... ideas about customer-focused organizations [sic], emphasises results over inputs and processes and implies that what citizens value cannot be presumed by professional communities in government. (1992: 119)

Public authorities, in other words, must become much more like businesses. They must become 'customer focussed', not serving the general interest of all citizens as expressed through the ballot box but somehow using other means to interact directly with citizens who express their wishes through a kind of individual conversation with the providers of public services.

This kind of rhetoric and rethinking of what is meant by the functions of government and their legitimation has created a serious set of tensions within governments and their administrations about both the direction of policies and their implementation. Into this tense arena have come the powers of the new communication technologies which can enable important changes to the relation-ships between governing and governed and to the administration of public programs. It is with the ways in which citizen–government tensions are being played out in conjunction with the new technologies that the first part of this chapter is principally concerned.

In most countries, despite what was said above, citizens have not altogether lost confidence in the capacity of the public sector to protect their interests and promote the common good. Since much public confidence relies on the use of the 'rational' knowledge associated with science and technology, science, technology and government remain three of the most important terms in the lexicon of societies at the beginning of the twenty-first century.

These terms together still invoke images of economic development and the generation of knowledge to stem public unease about such areas as problems of the environment created by unbridled economic growth. They also invoke the provision of services to a population and the arbitration of social claims and disputes in a 'rational' manner. Scientists, technologists and governments are expected in the popular mind to 'do something' about problems while also delivering just desserts to different groups of citizens.

People also understand that the three 'actors' – science, technology and government – represent power. They realize that the knowledge generated by science can underwrite socio-economic power, that technologies create systems which can serve the powerful, and that governments hold power. The question is, 'How much power and is it legitimately held?' The second part of the chapter looks at science for policy and policies for science and the role of science in legitimating policy directions.

Overall, this chapter considers some recent developments in the use of science and technology in public administration. The chapter defines 'public administration' as embracing all stages of the policy-making process. Put simply, the policy process includes the following stages:

- Agenda setting
- Development of policy options
- Selection among competing options
- Implementation
- Monitoring and evaluation
- Policy legitimation or change.

These stages are not necessarily sequential. In practice, many occur together, options selected continuously evolve as circumstances change, policy is made even during the implementation period and feedback occurs at all points. None the less, the framework provides a useful structure within which to organize discussion.

The chapter uses this organizing framework to discuss issues relating to power, especially power issues surrounding information technologies used in policy implementation and the additional 'surveillance' capacity information technology (IT) gives government.

The issues discussed also concern the 'leakage' of power that IT places in the hands of citizens which enables them to 'resist' if they so wish. Some observers have even opined that IT can change rules of the game, allowing new players to enter and change the rules, the game itself and the outcomes. For reasons of space, this issue is addressed only obliquely here but the theme of power and 'counter-power' runs through much of the discussion. Recent developments have caused many citizens to worry about the new technological capacities, which allow governments to chip away at the privacy of much of what citizens do, and thereby at their rights. Many share concerns that scientific advice given to governments may be too narrow and not consider the interests of the broader community. Citizen concerns are especially clear in areas where policy arenas overlap, coordination of policy is needed and where governments are often perceived to fail.

Computerization has promised many benefits to different areas of government. Operations research, for example, can be used in urban planning to indicate where to place hospitals, the implications for such decisions for transport systems, how the placing of dams for electricity generation will affect recreational use of areas and so on. Using the tools of operations research can help build different scenarios early in departmental policy considerations so the full budgetary implications can be worked out and a fuller set of options can be discussed. In this way, computer-based tools can improve the 'rationality' of public decision-making by providing better information. It can therefore assist at the policy stages of consideration of options. Equally important, however, the apparent neutrality of such tools may be critical to the legitimation of the decisions taken and to their justification in the broader public arena.

To some extent the outcome of the introduction of computer-based technologies depends on the technological competence of decision-makers. Governments, in the sense of elected policy-makers, are, however, not bodies of experts in anything except politics and have to seek advice on options both made available to them and almost pushed on them by developments outside government or by the complexities arising from the huge scale of operation of public decision-making and administration in most OECD nations. These enquiries are usually entrusted to specific expert bodies which are arms of government (specialist bureaux of research, for instance) or outside consultants.

Herein, lies the problem. Many choices that politicians and administrators make in the IT field are influenced by advisers who assume that the new capacity is to be used for control rather than facilitation of access by 'consumers' of government services. Outside experts who advise and design systems are from the private sector, hold private sector views and command private sector technologies. Most importantly, they make private sector assumptions about the nature of the 'market' to be served. The bias of such inputs may go unrecognized by the leaders of the administrative or program areas concerned.

This 'blindness' is partly a function of broader trends within the public service worldwide. The policy expertise of public officials has been greatly undermined in recent years in most OECD countries by the shift to the view that good managers can manage any system and any portfolio after minimum introduction to the substance of the field.

This shift has had serious consequences. In the fast-moving advanced technology fields of telecommunications for example, recent decisions in Australia indicate that senior departmental watchdogs with formal responsibility for policies adoption had little or no expertise. This, some have argued, has meant that the Minister's advisers were unable to brief him adequately on the full consequences and implications of the options considered (Joseph, 1993). Such low technical competence in the face of complex issues, which need urgent resolution, means a real loss of the power of government to act in the public interest while increasing that of dominant private players. This loss is especially serious in the face of the technologies which are *process* as well as produce oriented and which may be seen as the generic technologies which will underpin most governmental and administrative activities in the coming decades. Lack of expertise inside government not only risks capture of decision-makers by outside vested interests but also makes officials less capable of judging the competing claims of outside providers as to the accuracy of the technical claims made and the relevance of the claims in relation to the protection and promotion of the public interest.

A further reason why governments have not taken full advantage of IT's potential for citizen empowerment is the enormous investment in hardware, software and 'humanware' needed to raise the skill levels of users in the community and provide them with the necessary access to machines. Most current holders of formal political power have not made such investments. In America, for example, it has been left to community groups and the lower levels of government to design experiments involving greater user participation in public decisions and the design of policies. In other countries, as in the UK example discussed below, investment has been made instead in increasing the surveillance capacity of program operation. Sometimes this has been overt, as in the search for social security fraud in many countries; sometimes it has been less public, as in the linking of government departments' information into a single computer base.

Administrative technologies: retooling decision-making

In 1989 Stewart Clegg wrote that 'there is no such thing as a single all-embracing concept of power *per se*' (1989: 1). I would add that there is probably no such unitary notion of public administration. Public administration is part of the formal political process of a nation or sub-national unit.

There are many technologies which could be discussed in relation to public administration. The most pervasive is IT, now being used to an ever-increasing extent both systematically within the organizations of public administration and within the organizations of politics and policy development. IT technologies seem to be being used in a manner which suggests the considerable reshaping of power relations as between citizens and government. If organization is power, the information which 'escapes' from the formal administrative system and the technical capacities of the technologies themselves may be used to wrest power from some in formal power positions. The same technologies, however, increase the surveillance power of government.

IT-based media differ from others in providing for a greatly increased volume of information, increased speed of information transmission, greater consumer control over the media, increased capacity for senders to target messages to selected audiences and a technological push toward decentralization (Abramson *et al.*, 1988). Finally and critically, the new media provide the capacity for greater sender-receiver interaction. They permit both synchronous and asynchronous interaction. They allow one-to-one communications but also one-to-many, many-to-one and many-to-many.

Most research into the ways in which community groups are affecting public awareness of political and social issues and raising the stakes in some communities' interaction with government comes from the US. Most concern has been with citizen–policy-makers interaction at policy selection and adoption stage. Both the experiments and the evaluation approach owe much to a fairly specifically American belief in the value of particular forms of community participation in politics and public administration, such as those of the 'town meeting'. Most are concerned with grass roots consciousness-raising about issues of close concern to that area or group, not with changing major aspects of the system of public administration. Thus, while use of IT by such groups may increase the information available to particular sets of citizens, it has only incidental system effects. The examples given here do not relate to the much more institutionally 'political' side of the use of new technologies as in 'push-button referenda' (see e.g. Abramson *et al.*, 1988).

Facilitating consumer access to policy creation

Much policy-making in western political systems depends on input from public officials who work with ministers to create or modify a piece of legislation which will pass the political test (see Abramson *et al.*, 1988; Arterton, 1987; Bonchek, 1995). In some American states, however, politicians are bypassing both the experts and their own political colleagues, going directly to the public when options are canvassed and selections are made. By the mid-1980s, for example, eleven State governments had begun to distribute public information through electronic means and to provide opportunities for citizen input to policy proposals. In Alaska, for example, in fifteen locations Legislative Information Offices permit citizens to examine computer records containing the text of Bills under consideration in the State Legislature, providing the calendar for when committees, sub-committees and the Chambers consider legislation and indicate the voting records of elected representatives. From remote locations citizens can also voice their opinions to legislators over a free electronic mail system and a permanent teleconferencing network. (This and the following examples in this and the next sections are largely taken from Abramson *et al.*, 1988 and Bonchek, 1995.)

With similar concerns, in the mid-1980s a Michigan State Senator established two computer bulletin boards to allow his constituents and other citizens to provide input to policy-making. In 1985 he logged 10,000 calls from over 4000 people who left 12,000 messages. What he did with the information, however, is not clear.

The networks created by computer connections and information-sharing will alter the policy process. Research has investigated how and whether computer-facilitated voting from home on the 'town meeting' model, for instance in response to calls from politicians for expressions of opinion by constituents, makes for 'better' or 'worse' decisions (Arterton, 1987 and Bonchek, 1995). The results indicate that voting via home computer encourages the expression of more extreme views. Citizens voting from home lack group discussion of the crosscutting 'deals' essential to politics, leading them to vent their 'most immediate and raw responses' (Abramson *et al.*, 1988: 165).

Recognizing the problem, the Hawaiian televoting project in 1981 was specifically designed to provide voters with some civic education about the issues, which the Constitutional Convention was to address. Despite this effort, however, Abramson and colleagues concluded that the civic education accomplished by televote experience must be judged meagre (1988: 173).

At first sight, experiments with citizen participation on community development proposals in Reading, Pennsylvania in the late 1970s suggest that communications technologies can make a positive difference to how communities are governed. Reading was a pioneer in operating an interactive cable TV channel in the late 1970s. Popular demand transformed an information network for Reading's elderly population into a service for the general population and continued on the 'electronic town meeting' model. By 1983, Reading citizens participated in most hearings on budget and community development proposals. This experiment became the model for many discussions about using electronic means to increase citizen participation in public decisions.

Unfortunately, the Reading experiment was far from involving all citizens. Only 27 percent of the Reading population stated that they regularly watched the cable political programs. Abramson and others concluded, moreover, that even the relative success of Reading's experiment had not been duplicated elsewhere. In Dallas, for instance, a decade ago a panel of 'experts' was connected with four video-conferencing stations from which citizens could ask questions about planned mass transit schemes. Only around 100 of the city's almost one million inhabitants attended. The whole exercise lost credibility because the rules of questioning ruled out challenges and repeat questions and because the public had no input to the membership of the panel which was stacked with mass transit officials. Such obvious management of the event suggests that systems of electronic participation in decision-making by the governed are as open to manipulation by public officials as are non-electronic systems. This conclusion was confirmed by similar public teleconferencing arrangements in Alaska; ostensibly designed to overcome the problems of distance, they were packed by lobbyists for special interest groups (Abramson *et al.*, 1988: 184).

Improving service to clients?

Improvement of service to clients is a stated element in many plans for computerization, which does indeed work to client advantage in policy implementation in many countries. This is clear in some examples given below. In many cases, however, despite initial good intentions, cost-cutting concerns which co-exist with the other goals seem to impact in favor of the convenience of the administration when there is a choice to be made, especially in a welfare field. A study carried out in the UK a few years ago of the computerization of the operations of the Department of Social Security (DSS) showed this tendency clearly. As Helen Margetts points out in her prize-winning essay on the subject in 1990,

> The service has improved in some ways. For example, clients do not have to wait while case papers are found. However, most of the smaller projects that have disappeared from the plans were those that were to enhance the service to claimants, for example, VDU screens for claimants to work out their own entitlement before official assessment. [In addition] the accuracy of the new systems has been cause for public concern. (1991: 332)

After computerization it was largely 'business as usual' for claimants. Little progress in assisting the 'whole person' was made and claims for different benefits, while received at the same office point, continued to be processed and implemented separately. Staff within local offices still dealt with one particular benefit and there were no plans to train general benefit advisers in local offices. Claimants were not necessarily notified of other possible claims they might make. Management interests took precedence over all other elements of the justification for the 'largest civil computing project ever undertaken in the world' (Margetts, 1991: 331–3) while there was routinely failure to understand that the interests of job satisfaction for staff, client satisfaction and cost-saving may be internally contradictory.

Little evidence that any major transfer of knowledge from the private sector consultants used for the project to the DSS staff took place. This lack of skilling of internal staff meant that there was no one with the expertise to monitor the system or advise on how to do it better next time. The staff continued to be as dependent on the decisions of top management as they had been before the installation of the new system. Perhaps most telling in relation to the realities of the power mode governing the project is the fact that no users directly involved in dealing with clients were consulted about the design of the new system, despite such input by users being recognized as the key to success in computerization of administrative operations. As a result, for instance, crucial notepad facilities for staff dealing with clients were not made available on the computer and no one rethought the policy improvement possibilities that the new methods of work made possible. Even before the strategy was implemented, analysts were warning that:

> the overall tone of the Operational Strategy was aimed at how best to run the administrative machine . . . rather than how to serve the public or effect an anti-poverty policy. This is in stark contrast with the activities of German local health funds which, despite the constraints arising from centrally planned systems, have managed to link benefit systems with policy concerns and have developed a strategy for attracting more customers by improving service quality. (Avgerou, 1989, quoted Margetts, 1991: 336)

It had become increasingly apparent elsewhere that the new tools of government are value-laden and power-related instruments which influence orientations in and around the public administrative apparatus. Used in the implementation phase of public decisions, expert systems, for example, may cause problems by excluding potential claimants because computers require strict criteria for acceptance of claims, not allowing the use of common sense to decide marginal cases (Snellen *et al.*, 1989). Snellen and his colleagues demonstrate the issues by discussing the Norwegian system of payment of student support grants:

> Using the PIN (which each citizen has) the system accesses the central population register. It determines who is registered as inhabiting the same dwelling . . . . Access to the tax register allows the system to determine the joint income of the household and access to the National Insurance Institution gives information on the welfare related status of household members. This information then is added to that held by the state housing bank which adds information on the age of the housing unit, its location in relation to the centre of the town, interest to be paid on bank loans etc and calculates what may be due in housing assistance. The system itself then writes to the applicant to explain why the application has been refused or if the application has been successful sends a cheque. All without any human intervention. (1989: 7)

This is in many cases an efficient and equitable system. Its disadvantages derive from use of many different databases, not all containing current information, and from the interpretation of critical terms built into the system (Snellen *et al.*, 1989: 8).

Uneven data quality is a problem everywhere. Garbage in, means garbage out. A study of recorded criminal histories carried out in the early 1980s in North Carolina, for example, showed that only 12 percent were complete, accurate and unambiguous. Similarly low figures were found in California (19 percent) and Minnesota (49.5 percent) (quoted, Snellen *et al.*, 1989: 127).

Many further problems arise from the translation of legal ideas about entitlements into computer language. Computer languages cannot tolerate ambiguity. System designers thus make decisions about claimant eligibility immediately policy implementation begins, meaning that computer program designers make permanent decisions as to the law. The resulting system becomes a framework in which officials implement policy work, limiting them to the options set out by the designer (Bing, in Snellen *et al.*, 1989: 117–18). The term 'supporting spouse' may in practice involve diverse situations but these cannot be allowed for in the design of programs determining benefits so claimants may be wrongfully excluded. Even more seriously, problems of meaning of critical terms such as 'income' have been used to prosecute claimants. In a famous, or infamous, case in Sweden, for instance, it was ultimately found that the Swedish government used the term in more than 25 different ways (Bing, quoted in Snellen *et al.*, 1989: 126).

Surveillance

Paradoxically the load on governments as providers of services and the social wage almost always increases in times of recession since payment of unemployment benefits, administration of training programs, etc., are a vastly increased part of government activity and expenditure. These are costly to administer and the rules increase in complexity. There is a tendency then to try to cut costs, reduce staff, move to more rule-related and hence automatable eligibility assessments, all in the name of 'improving the service to the public'.

Into these circumstances come new technologies which increase the possibilities of civil surveillance by government and their officials. There has been much concern about the way in which the search by public administrators and their political masters for efficiency and cost-cutting has focussed on control of fraud. In Australia a decade ago, the federal government, seeking to reduce both taxes and social security fraud, proposed to introduce the Australia Card. This card was an electronically readable database, which included tax, social security and other personal data and could be read by a variety of public sector agencies. Its use meant that an agency which dealt with a citizen in one set of circumstances could easily obtain access to the same citizen's dealing with other agencies, thereby verifying the accuracy of the information provided, for instance in relation to income declarations. This card was defeated in Australia by a coalition of experts and community groups, although a universal tax file number system which works like the American Social Security number system means that citizens can be easily identified as they must provide a number when being paid many parts of the social wage, applying for a bank account and many other of the essential tasks of life. In principle there is no cross-checking by different agencies but it is not unknown for one public agency to sell information on clients to interested parties, as a recent case in New South Wales demonstrated.

The privacy of personal data is much less assured now than in the past. Studies of the data files that businesses keep on their workforces have indicated the scale of inaccuracies as well as the range of data stored. One such study of personal data management by seven large private sector organizations in the early 1990s showed that in many cases there was little or inadequate policy to determine what information was kept, the form in which it was kept and the accuracy of the data held (Smith, 1994). Smith also noted that even where policies did exist one could not assume that they were honored systematically. There is no obvious reason to expect that things are much better in public sector organizations anywhere in modern societies. While to some extent the rights of employees to their personal files may act as a check on accuracy, few individuals look at their files systematically. Freedom of Information legislation also seldom acts as total protection since files can in many cases be censored before being revealed to enquirers.

There are studies, too, which document the heavy pressures put on public and private organizations to reveal confidential information. This is often the case with medical data, especially in relation to infectious diseases such as AIDS. Smith's study in the US shows that traffic in 'confidential' data from private physicians and hospitals is widespread (1994).

In Australia, a recent committee set up to make recommendations about using the emerging broadband possibilities to 'make information technology a more efficient and effective means of achieving the objectives of government' made a remarkably positive set of recommendations. The Committee expressed great enthusiasm about the potential of the new technologies in public administration. The Committee said that 'the benefits of cross-agency uses of information technology need to be more vigorously pursued in the interests of greater efficiency and better client service'. It continued by suggesting that information technology staff should be regarded as a shared resource across the central government and their skills should reflect an understanding of the business needs of the public sector. Moreover, the Committee said, the option of outsourcing needs to be considered in terms of defining the core activities of agencies and in improving agency focus. Information systems developed for administrative functions within agencies should be rationalized to reduce duplication and provide greater cross-agency flexibility for program delivery. Since, the Committee continued, communications and computing are now inextricably linked, similar rationalization of data networks should also occur to achieve scale economies and operating efficiency. Privacy guidelines that were developed in an information systems environment dominated by central computing should be reviewed in the light of the report in order to take account of the potential benefits offered to clients and to the Commonwealth by new systems approaches (see Information Technology Review Group, 1995).

In making its recommendations, the Committee apparently assumed that the interests of efficient public administration and client service were identical. Its members took a highly sanguine view of the risks to personal privacy posed by their recommendations as to outsourcing, use of the same staff across the government and cross-agency use of data. Their propositions indicated that they considered that such risk to individuals in the reshaping of the power balance between information held by individuals and governments did not weigh as heavily as the convenience to and cost-cutting made possible for the government. That the Committee's recommendations refer specifically to government's weaker clients, such as jobseekers, welfare beneficiaries, hospital patients and school children, as well as more generally to all taxpayers makes this equation of efficiency and improvement to client services worthy of much greater scrutiny.

Abuse of the possibilities of crosschecking or the sale of information may perhaps be relatively rare. In many instances, however, the new technologies are legitimately, or rather legally, used in ways which are not public but which have the potential to damage the reputation of many innocent citizens.

This is shown by a recent study of the use of Geographic Information Systems (GIS) to underpin efforts to reduce social security payment fraud in Australia while apparently being focussed on improving service to clients through making urban planning decisions easier and more transparent. In 1994, a researcher at Deakin University published a paper which indicated how the proposed use of GIS in Victorian government agencies, especially local governments, raised important issues relating to privacy. He suggests that the use of GIS technology should be carefully monitored before this means of 'social surveillance becomes the new art form of planners of the modern Australian state' (O'Toole, 1994: 345).

The use of GIS in Victoria has been advocated for a variety of government–citizen interaction purposes because of the problems caused by the increasing use by policy-making and data on local populations drawn from diverse sources. GIS can put order into the confusing mass of data collected on different bases by using a computer system to collect, display, manipulate and integrate geographic and other data. With map coordinates, data coded to specific points can be displayed and various forms of data viewed from a single terminal. Local authorities already use Land Information Systems (LIS) to quickly review applications for building and see how these fit with other planning needs, provision of services, etc. GIS goes beyond this to permit the use of *individual client addresses* as boundaries for the provision of specific services or addressing the needs of particular ogranizations.

It is at the level of data matching that the problems of power in governance arise. These problems can go right to the heart of our democratic and judicial expectations. One of the dangers of computer matching, O'Toole points out, is the conversion of a presumption of innocence into a presumption of guilt. He quotes a former Canadian Privacy Commissioner as saying that:

> a computer match is instigated not because a particular person is suspected of fraud – as in a traditional investigation – but because a whole class or group of persons has come to the attention of government for either good or frivolous reasons. Thus do old-fashioned 'fishing expeditions' pose as high technology. (Grace, cited in Greenleaf, 1991 and recited by O'Toole, 1994: 348)

Greenleaf also cites the Australian Privacy Commissioner, Kevin O'Connor, as likening data-matching to 'the privacy equivalent of drift net fishing'. O'Toole concludes that with the current limitations of the Victorian Privacy Act, GIS using point data for individual households means that local residents are in real danger of invasion of privacy. Although O'Toole does not give precise examples, the privacy problem may arise because, using GIS, a social security agent could use data on established fraud patterns and a whole series of data not normally gathered by social security officials to tar all members of the population in particular geographical areas with the same characteristics with the same brush. This identification can be followed by officials subsequently using the frequency of such characteristics to increase surveillance of the whole population living there, whether individual households show any indication of fraud or not. This means that all households who live close to populations suspected or condemned for fraud in a given administrative portfolio could easily be subject to the same extra scrutiny.

Data-matching is a major problem for persons interested in social justice everywhere because use of computers to detect fraud has usually been aimed at the poor. In a powerful book on the dangers of computerization in government administration, Burnham has shown that the all-pervasive computer-matching of data under the Reagan administration was justified because it would 'save the taxpayers millions of dollars' (1983: 28). At the same time, however, the Labor Department acknowledged that making databases 'clean and accurate is very difficult. Social Security, for example, is a real mess' (quoted Burnham, 1983: 28).

IT can be used at the monitoring stage of the policy process to check on how 'horizontally' similar administrations are interpreting legislation and other kinds of public decisions. Using IT, each administration knows exactly what each of the others is doing and can coordinate action, effectively creating one system whereas before there had been several. One example of this new checking potential is the computer reporting system set up in the offices of US attorneys across the country which allows instant access to recent decisions and actions taken in the domain of legal actions. The new reporting system has meant that the Attorney General's office has subtly altered the process of justice administration by centralizing information on cases and precedents in every State.

A new computer database used in Congress now provides a check on the administration of federal largesse. The system provides 'geographic profiles' of federal grants and contract dollars allocated to congressional districts, enabling Representatives to compare the distributions to their constituencies with allocations to other areas and over time. This enables elected representatives to keep a continuous and easy check on the activities of the administrators charged with implementing Federal programs, thus subtly changing the balance of power by enabling complaints about that administration to be made with clear data to back them. In this case, IT increases the power of elected representatives over public servants, whether elected or career.

Other databases are again becoming a permanent part of the political and administrative landscape and altering the operation of the fields in which they are used, particularly in relation to the 'horizontally' equivalent administrations below central government level. In addition to bridging the gaps between State administrations and elected officials of America's major cities to provide instant information on policies and issues of the day. In this way, to the extent that the problems of federalism are born of ignorance or the inability of political authorities to communicate with each other, the emerging media will ease the problems of coordination across branches of government and across levels of the political system. There is already ample evidence that such media will allow political elites to access the information they need about the actions of other political authorities. We have seen from the examples given above that the new media also allow the same sharing of information among the administrators implementing and monitoring the decisions made by their political masters.

Information for democracy?

If rapid and reliable access to information is critical to the proper functioning of democratic processes, who will provide that information and what will be its content? The market place does not seem likely to support electronic information services devoted to exchanges of civic information. Mass programming dominates the cable channels' offerings. Films dominate, followed by sports, music, concerts, etc., with politics running a poor last. Moreover, as cable subscriptions show signs of slowing their growth the cable companies have moved toward pay-per-viewing methods for programs, which emphasize conventional entertainment. The most successful viedeotex services are devoted to narrowly economic information.

Perhaps, however, the Internet itself can overcome some of these problems since it costs very little to subscribe and there is virtually no censorship of the material available? The Net has become a medium where individuals of many persuasions and organizations with many agendas regularly put out information ranging from the scholarly to the pornographic. Much of this is 'news' of one kind or another and indeed most of the major newspapers around the world can now be accessed via the Web, as can some of the resources of major libraries. By the 1990s, young people in most OECD countries were gaining familiarity with electronic means of communication at school, and universities are now moving to require all students to work with electronic data sources as well as in the traditional classroom. Although PC prices are falling, however, they are still certainly not within the reach of every family for every member (one in five Australians now has access to a computer on a regular basis) and most Internet users are young, single and male.

Whether Internet-disseminated information assists democracy or not depends on both the material which is available on the Net and on the questions which young people are being trained to ask. If people are not endowed with basic political capability through their education, are not trained to be questioning in relation to matters of civic concern, they will not be able to ask the questions which exercising their democratic right involves. If, for example, they are not taught that they can participate in administrative decisions which affect them locally by taking part in public hearings and making submissions to governments, they will not do so. Instead they will remain passive in the face of the information put out to them via the new media just as they do in relation to the traditional means of communication. If they are not taught that administrators can be criticized, that it is OK to blow whistles when the occasion demands it, if indeed they are not taught how to judge when the occasion demands it, they will not use the latent power they have. It is not enough for them to know their rights in an abstract manner. Making submissions demands practice. Being effective in the public arena has to be learned. Underlying skills have to be learned at school. Newspapers in modern western societies are increasingly bland and do not present as many alternative opinions as they did in the nineteenth century, let alone encourage citizens to act on their beliefs outside the ballot box. IT alone cannot compensate since information needs a framework to become knowledge.

Experiments such as that recently carried out in Victoria, Australia, which used inserts in newspapers to ask readers whether sentences for a named set of crimes should be increased, are an exercise in public manipulation. Civic education in Australia, as was recently pointed out by the committee of inquiry chaired by Professor Stuart Macintyre, is lamentably inadequate and does not allow citizens to make informed judgments on emerging and complex issues (Macintyre 1994). Having access to the Net in these circumstances is of little value in keeping public administrators in check and politicians 'honest'.

The increased sway of principles of 'user pays' is also threatening to make it harder for citizens to acquire information which has been collected with taxpayers' money. In a paper written in 1989 Macdonald and Reams point out that much information is now only available to the public at a sometimes quite considerable charge. In Australia this is true for much Australian Bureau of Statistics data, for example. Macdonald and Reams remark that:

In as much as democracy and the rule of law require an informed citizenry, they are threatened by charges for government information, whether they are levied by the private sector or by government itself. Following the adage that knowledge is power, political, social and economic power are likely to go to those who can pay for information. (1989: B16)

Resistance

Instances of resistance by consumers to the increased concentration of power made possible by the introduction of IT to the administrative systems of government are becoming more common. Weatherly has shown, for example, that clients in many program areas have the potential to defraud the system, despite the introduction of more and more computer checks, but, perhaps surprisingly, they tend to comply, driven by notions relating to their perceptions of the justice of the system as a whole. That may change with extra surveillance and become a sentiment of injustice, giving people a 'moral' right to exercise their talents in defrauding the government. In some cases, the IT which they in turn have at their disposal may work in their favor. It is already clear that, for whatever reasons, the Internet provides many possibilities of taxation fraud (*Australian Financial Review*, 3.2.96, p.1). These questions remain understudied for the moment, both in terms of the possibilities of resistance and in terms of the motivation for such resistance which may be critical to achieving shifts in the balance of administrative power.

Developing knowledge for policy: science in the new socio-political order

Most OECD governments have come to rely on their successful management of the economy as the basis for their retention of political power and as the mechanism whereby they can 'afford' to provide the many social services and protections against adversity which their citizens have come to expect. The basis for economic growth has therefore taken center stage while in more recent years the market has been assigned a greater role in achieving desired outcomes. Most economic advice to governments over recent decades has come from 'experts' trained in neo-classical economics. It is their prescriptions which have underlain the cost-cutting and reduction of government framework discussed earlier and which impinges on the implementation aspects of government especially.

The focus recommended has not always provided easy solutions, however, and policy-makers in some OECD countries at least, as well as the EU itself, have come to recognize that markets may need to be assisted by interventions designed to lift scientific and technological and hence industrial performance. This means that they are beginning to seek the advice of a new group of 'experts'. This section looks particularly at some of these trends and their derivation.

In recent years increasing policy attention has turned to the analyses of economists using the legacy of Schumpeter to demonstrate that technological change is central to economic development. Recognition of the importance of R&D and innovation as engines of growth and the improvement of national competitiveness in the emerging 'learning economies' of advanced nations has been expressed in many policy statements by OECD member nations. For their part, the Newly Industrializing

Economies (NIEs) of Asia have made technology development and transfer the keystones of their industrialization policies (see for example, Tyson, 1992; Wade, 1990; Wong, 1995).

Attention to technological upgrading as the basis of competition has increased as international agreements and national political leaders have encouraged trade liberalization policies and moved toward deregulation, corporatization and privatization of key sectors. These trends have made it imperative that manufacturing and services move toward the leading edges in their fields so as to win in the international export races which now dominate government attention. This new economic imperative in turn has coincided with the growth of major national and international problems, the solution of which needs greater scientific input to policy development. Central here are the complex policy arenas related to protection of the environment and sustainable development. The findings of scientists are therefore critical to policy on many fronts.

Policies for science

Together, the trends discussed above have placed greater public demands on the capacity of science to contribute to national economic development while pushing the funding of R&D towards the private sector. In the 1980s, in most OECD nations, new public policies were put in place which encouraged the private sector to invest more in R&D to a greater extent than they had in the past, especially in countries such as Australia where business investment had been notoriously low by OECD standards. The result of these policies was to increase the overall proportion of R&D carried out by the private sector.

By the beginning of the 1990s, however, the proportions of national R&D contributed by industry were beginning to drop again in many OECD countries. Hit by depressions and a general decline in the business environment, many major companies preferred restructuring their operations to investing in the riskier areas of their business.

It seems that several trend responses by governments to this decline can be seen. First, despite pressures to cut overall public spending, many governments are increasing public funding for basic research. The watershed years seem to have been 1994 to 1996. Japan decided in late 1995 to increase investment in basic research, especially in advanced scientific fields, lifting its national science and technology budget for 1996 by 6.9 percent over 1995. In Germany, increased support of 5 percent per year for five years was promised for university centers of research excellence. In the US, a slightly different path was followed. There the government invested in the creation of facilities, which would serve the needs of government, university and industry research communities. This investment raised the proportion of public investment in basic research from 13 percent in the mid-1980s to 18 percent in 1994.

In some countries, basic, mission and applied science have all benefitted from increased government spending on science. In 1996, the Netherlands budget, for example, recognized the need to both broaden and strengthen the level of public support for scientific and technological research. Most of the Nordic nations also increased their public sector science budgets, Denmark planning to reach the levels of expenditure of its neighbors before the year 2000. Similarly, France increased public funding for R&D and innovation by several percent in 1994 and 1995.

At the same time, governments are beginning to make difficult choices. In recognition of the high cost of basic research and of the fact that small countries in particular cannot hope to compete or even to contribute significantly to science development on all fronts, public support is being concentrated in key sectors. Most such sectors involve several disciplines in the natural sciences and should involve more than they do in the disciplines of the social sciences. Sectors selected usually include information technology, energy, technologies connected with environmental issues, biotechnology and health, advanced materials and in some cases space and aeronautics. IT, biotechnology and advanced materials are core technologies, already identified by observers in the late 1980s (see, for instance, van Tulder and Junne, 1988), the importance of which many governments have only recently recognized in funding allocations.

This example shows how cautious most governments are in making decisions about the priority areas to fund. Perhaps the allocation of resources to the sciences relating to the environment is the classic case of slow public sector response to emerging problems and perhaps also of slow recognition by researchers themselves of trends affecting their research directions. This will be discussed further below when discussing scientific advice to governments. It is clear, for example, that the problem of global warming has been known to the scientific community now for nearly half a century (Hart and Victor, 1993). That major areas of funding should only now be diverted to research on environmental concerns of all types should concern all scientists with an interest in policy.

Recent shifts in government attitudes to funding and to priority may be related to trends outlined in an influential work by a group of internationally renowned science policy analysts. This work, published as *The New Production of Knowledge* in 1994, pointed out that in many fields the frontiers of science are now being pushed back by scientists working in arenas financed and organized in very different ways from those traditionally associated with scientific endeavour (Gibbons *et al.*, 1994). The book suggests that a new mode of knowledge production is emerging (Mode 2). Whereas traditionally scientists worked in single discipline teams, financed by universities or public sector laboratories, under the direction of long-established academic experts whose work is judged uniquely by peers (Mode 1), many of the socially and econo-mically most important results are now being generated in the context of application by transient, interdisciplinary teams who undertake tasks defined and commissioned by agencies which have to deal with areas which defy the 'normal' approaches.

Knowledge produced through Mode 2 arrangements may be commissioned by environmental protection agencies, by community groups, by government bodies themselves and by industry or groups of companies. Once their task is completed the teams working on the issue may disperse but remain in contact via new means of communication and create networks of researchers who are not subject to the rules and conventions of academic science within the universities.

Slow government recognition of the shift which is not reflected in resource allocation decisions means that much new knowledge is being generated haphazardly and may never be published in the normal manner. The new knowledge is thus not subject to peer review, a limited but important check on validity and reliability, and never becomes part of the background knowledge taken for granted by scientists working either in the more conventional university arenas or in government.

The development of Mode 2 knowledge production has been made possible, suggest Gibbons and his colleagues, because most of the basic principles of science have been discovered. What is now of most scientific concern is the application of existing knowledge; less the isolation of new molecules, for instance, than the knitting together of these basic elements into new drugs. This trend also means that scientific information generated by the public sector is increasingly of direct value to the industrial sector while it is also a tool for action by groups concerned with emerging issues in arenas where action was once considered to be the domain of public authorities.

Perhaps the most important of these public issues is the environment, where many groups feel both that governments provide insufficient funding through the allocation procedures of Mode 1 and neglect critical issues. The development of research through Mode 2 mechanisms is pushing scientific endeavor in new directions and making the distinction between basic and applied science more difficult to sustain; it is also pushing the contribution of science to policy debates into new areas and involving new players in policy and social struggles of all kinds. This trend may both make science much less mysterious to the general community, as more groups commission and are able to use its results, and greatly influence notions of what constitutes 'good science'. The trend may also increase the problems of *misuse* of scientific results if results have to be greatly simplified to allow more general public participation in policy-making, especially using the electronic decision technologies which were discussed in the first part of this chapter.

Bringing together the interests of universities and the private sector and linking these with the concerns of public authorities is also high on government agendas. This concern has translated into policies linking funding of science to new organizational arrangements for its development, notably arrangements in which industry and public sector research (PSR) work together. These include in Australia the program funding 62 Cooperative Research Centers which link industry and PSR institutions in research related to the development of new products which will become proprietary. In other countries, such as Denmark, this linkage effort has been more focussed on the joint development of technologies with wide applicability.

Given the high cost and risky nature of investment in basic research and the recognition that much more knowledge of use to industry and government is now available than is known to most players, many OECD governments are increasingly putting resources into the *diffusion* of existing information. Knowledge diffusion is seen as a direct contribution to improving the productivity of existing plant and skills. It is a relatively low-cost means of fostering productivity, economic growth and international competitiveness by increased use of available best practice as demonstrated in programs such as 'Building and Living' in Germany and US industry-led initiatives including the Advanced Technology Program and the Technology Reinvestment Project.

In several countries new organizational arrangements for knowledge diffusion have been created. In Sweden, for instance, Competence Centers affiliated to universities have been established to serve as a strategic resource for the technological renewal of industry, an aim considered important despite the fact that Swedish industry has both performed and funded higher proportions of R&D than in most other OECD nations. Norway has other schemes designed especially to assist technology to diffuse from universities to small private firms. Canada, too, has been active in this field, the Canadian Technology Network providing firms with advice and easy access to PSR knowledge and researchers. Most OECD countries also run technology consultancy services, notably for small and medium-sized enterprises (SMEs), with knowledge diffusion rather than production in mind. A recent book by Bessant and Dodgson (1996) outlines many such schemes throughout Europe and in Japan.

As the need for development and adoption of new technologies has become more apparent to governments a wide variety of consultative and advisory bodies has been set up in all industrialized countries. In the US, the President has his Science and Technology Advisory Council, as does the Prime Minister of Australia. The German Chancellor has his Council for Research, Technology and Innovation and the Japanese Prime Minister is directly advised by his Council for Science and Technology. Until recently, the American Congress had its own Office of Technology Assessment, a notion copied later by the Netherlands which set up NOTA (Netherlands Office of Technology Assessment) with the same brief. Similarly, Canada has a National Advisory Board on Science and Technology and the Council of Science and Technology Advisors while Finland has a Technology Policy Council. Behind these formal bodies lies an enormous array of further committees and agencies which all have their own scientific councils or panels of advisers.

National systems of innovation

Recent recognition by governments of critical findings of international research has also affected the shape and focus of public policies linking science and technology to economic development goals. These findings, the legitimacy of which have been assured by a series of projects carried out by the OECD, point to the importance for economic growth and competitiveness of the functioning of national systems of innovation (NIS). The research suggests that a country's development can be maximized only if that nation's system of innovation is working efficiently.

A national system of innovation comprises institutions and links between institutions that influence patterns of information-flow between producers and users of economically useful knowledge, both formalized and tacit, embodied and potential. The system is composed of education (especially colleges and universities), training organizations, regulatory and policy-making institutions and businesses themselves and includes specific trade and development support organizations.

It is also composed of institutions in the economists' sense of the 'rules of the game'. These rules include the nature and adequacy of legal frameworks governing operations, such as those relating to intellectual property, rules governing competition (such as anti-trust laws which make it difficult in some cases for enterprises to invest in joint marketing, R&D, etc.) and the regulation of public sector purchasing and labor markets. A NIS is also composed of regulations which affect such areas as the environment which influence the ways in which businesses can compete (produce). A properly functioning national system of innovation may be defined as one whose different aspects work harmoniously and there are no serious differences of policy or operational goal which reduce the efficiency and effectiveness of the functioning of major elements such as the PSR and business taxation systems.

Public policies increasingly aim to harness the power of all elements of national systems of innovation, explicitly in some countries, more implicitly in others. Countries which use their rule-making powers in ways which do not have this focus may be wasting resources and not receiving maximum benefit in terms of national economic development. The ways in which governments perceive the functioning of their economies therefore has particular and increasing salience. It seems that while there is a growing trend for the public sector to undertake fewer functions than the norm since the end of World War II and to play a less direct development role itself, smart countries are none the less recognizing that the market alone cannot always be effective without particular stimulus or new types of regulation, without policies to develop leading edge technologies and diffuse them as widely as possible. This view is evident in the research-funding policies described above.

Making these policies effective, however, depends on a sophisticated understanding of what stimuli will work and where to draw the line between letting industry manage and providing assistance to move in the right direction. Many countries have recognized the role of new knowledge in the emerging learning economy. Most mechanisms to encourage greater investment by industry involve modifications of the tax system and subsidies. These may include tax breaks for new investment, provision of science and technology parks and other forms of encouragement to companies to move locations or find new export markets. Recent trends suggest a decline in the use and effectiveness of such incentives, although many countries still use 100 to 150 percent tax reductions for the performance of R&D.

Such policies, however, may be low impact because they fail to take account of a further trend in the organization of business itself and a more productive direct relationship with the public sector, both in its research arm and its policy-making sections. Two sets of approaches are discussed here.

The first concerns the provision of technological infrastructure (TI) and the associated Technology Infrastructure Policy (TIP) discussed by Justman, Teubal and colleagues in *Technological Infrastructure Policy: An International Perspective* published in 1996. As Justman and Teubal say,

> Accelerated technological development and the globalization of trade and investment have ... [increased] the importance of technological capabilities as a source of competitive advantage ... [and] raised new needs and new opportunities for collective action in support of individual firms' efforts to acquire the necessary capabilities. We refer to the public goods that are the object of such collective efforts as *technological infrastructure* and to the policies aimed at promoting their creation or emergence as *technological infrastructure policy*. (1996: 21)

Technological infrastructure is less tangible than the conventionally recognized infrastructure of roads, rail, telecommunications systems and 'often involves the articulation of new needs that can only be met through the generation of new capabilities within markets that have yet to be created'.

Public authorities find it hard to deal with the new needs because of the indivisibility of infrastructure and interdependence with private investment in firm-based capabilities and the difficulty of a context of limited appropriability, incomplete information, strong differentiation and rapid change.

Justman and Teubal therefore advocate new approaches to technology policy, placing greater emphasis on institutional innovation rather than price-based measures that compensate for the external benefits of technological change and viewing the public sector as essentially catalytic. Importantly, they emphasize, broadbased subsidies or tax credits cannot resolve the complex dimensions of coordination and public choice involved in building advanced technological infrastructure.

Governments must perhaps become more proactive in developing technology than they have usually been in OECD countries in the past and develop innovative mechanisms of policy advice for science, technology, industry and research. This may mean changes to traditional science advisory mechanisms and in the composition of skills and expertise which the public sector should expect to have at its disposition either inside the permanent public service or at its elbow.

The distinction between basic and advanced technological infrastructure suggests that there are different roles for government to play in developing the two areas. In establishing basic TI the government can play an active intermediary role that complements the role of the private sector in importing new technology. In facilitating the production of advanced TI, however, governments can do no more than facilitate private sector efforts to create new technology.

Importantly, successful technology infrastructure policy will require institutional innovation so that the government can be an effective interface between professional analysts and political forces. Justman and Teubal suggest some models which include MITI, an independent Technology Council acting outside the normal channels of government and depending for its authority on the professional and personal standing of its leaders or a tripartite commission composed of government, industry and labor representatives or an inter-ministerial committee. The organization selected must be a clearing house for information, scanning the horizon and searching for potential projects. Critically, to be successful in developing TIP, the government must have sufficient capability itself to recognize opportunities and deal with them appropriately. This echoes a point made earlier in this chapter about the vexed question of public sector capability in assessing the information technologies they use in new modes of public administration.

All these suggestions fly in the face of current trends in many countries, especially those where more conservative governments are trying to reduce the capability of the public service. Such governments believe that their sole role is to remove impediments to business activity which they view crudely as excessive taxation, wages, paperwork and bureaucracy. They understand little of the international shifts in the bases of economic growth and even less about technological change or paradigm shifts in the production field.

Such governments view their nations' economic activities as created by individual actors called firms, not by groups of enterprises. They are ideologically opposed to collective activities and do not understand that in the new environment firms must collaborate as much as they compete.

This blinds them to what Rycroft and Kash (1999) say lies behind successful science and technology policy in the 1990s and beyond. Rycroft and Kash emphasize that governments must recognize that successful continuous improvement of complex technologies depends on the development and maintenance of 'technology-based communities' because many technologies are too complex to be individual creations. These 'technology-based communities' are groups which cumulatively acquire a seamless mixture of cognitive and tacit knowledge and capability. The organizational structure needed for devising and implementing policies for technological development is therefore the network, not the firm or the agency. Its organizing locus will be the technological product/process sector. Its internal processes will be governed more by cooperation than by competition and its decisions will emphasize consensus not conflict. In sum, say Rycroft and Kash, 'continuous innovation of complex technologies appears heavily dependent on communitarianism' (1999: 615).

To understand what is involved in practice the authors reanalyze the linear model of innovation itself, pointing out, as others have done before, that the model only works for a few simple products such as petrochemicals. Importantly, they say that the model conformed to the structure of innovation in the mass production industrial system and that the dominance of the model in thinking about and policies for innovation in the US 'doubtless resulted from its convenient fit with American political/economic ideology. This ideological dimension can be seen in the way in which conceptualisation about the role of government in generating innovation conformed to a clear public–private distinction'. As they suggest,

> between the two sectors at the dividing point between R&D (activities which suffer from the 'free rider problem') and commercial activities, that is between pre-competitive and competitive activities. This dividing line has been ideologically convenient because it defined where government intervention and support were legitimate and where they were illegitimate. (1999: 617)

They point out that the linear model has been especially useful because it allows the dividing point to be moved in response to immediate political pressures through redefining what is covered by R&D. A key myth was that the market hand will efficiently move ideas and methods produced by R&D into commercial products and processes. In contrast, in reality, it is often difficult to work out where the ideas used in incremental innovation, the most common form, come from, especially where complex technologies are concerned, and we now know that the sources vary as technologies develop (see for example, Faulkner and Senker, 1994 and 1995).

Rycroft and Kash further suggest that successful networks cross both public and private sectors in new ways. They believe that it is useful to see them as linked into what they call 'complexes' which they describe as 'holding companies for multiple networks, organisations and individuals' (1999: 620). The examples they use in the US include agriculture, defense/aerospace and medicine. These fields, they say, are the success stories of American technological development and the source of most of the technologies that produce large trade surpluses. In Europe the EU has also been developing such complexes in the areas deemed strategic.

Extending the success of these complexes may mean that policy-makers realize that they must deal with three intersecting systems: systems of technological artefacts; the socio-technical systems which innovate and produce the artefacts; the complexes that act as holding companies for the development of major areas of technology. These intersections mean that the existing rigid distinction between public and private spheres should not be maintained, pushing governments to reconsider means of devising and implementing policies for economic development and other areas where science and technology play important roles in solving problems.

Science for policy

Making the adjustments to the policy development, implementation and monitoring process referred to above will not be easy. The use of science for policy in areas outside the quest for economic development is a highly contested domain.

This contestation occurs despite the general acceptance by lay persons, politicians and the scientific community alike that increasing the influence of scientists in the policy process is important in removing decisions which affect the public interest from the corrupting influence of the particular and partial interests which dominate the domain generally defined as that of 'politics'.

Science is often, but erroneously, thought to be 'above politics', to be a neutral set of ideas, theories and facts which can be used to settle partisan disputes and reduce social conflict. This view assumes that disputants will see the truth in the light shed by scientific evidence, enabling them to agree on desirable outcomes and policy directions.

The view of science as a neutral belief system allowing adjudication on matters of fact further supposes that people are rational actors who have only to be educated by scientists to abandon the 'unreasonable' positions they initially held to. The view further assumes that as people see that policies are conceived, implemented and monitored in a rational manner the decisions taken will be uncontestable and enjoy increased public legitimacy.

Unfortunately, it is now clear that even large-scale inputs of science do not seem to reduce the degree to which policy-making remains a question of 'muddling through', compromising and seeking consensus on solutions to issues to which there is no one right answer because the problems arise because of clashes of value. To see science and scientific evidence as either 'right' or 'wrong', correct or incorrect, is to be confused on both the nature of scientific endeavor and its power in an inevitably 'political' context. Most of these confusions are well known and widely discussed but bear some elaboration here.

Governments tend to call on scientists for assistance with the most complex and difficult issues which cross a wide range of public domains. In these areas governments are constantly tempted to call on science for assistance at *all* stages of the policy process. In particular, the voices of scientists help define political agendas in many complex areas of human activity which impinge on the natural world.

Unfortunately, these are precisely the arenas where the complexities begin at the level of even defining what the problem are, let alone finding solutions to them. How are politicians to recognize when environmental concerns are 'justified'? This relatively simple sounding question involves problems of definition of what constitutes a 'healthy' natural environment before one can begin assessing apparent damage and its usually multiple causes. The complexities continue when assigning causal priority to certain factors to indicate priorities for policy attention and legitimate the allocation of scarce funding resources. Even more difficult is the process of deciding how to monitor and evaluate the success or failure of the policy. It is especially hard to choose the criteria for deciding when a policy can be sensibly said to have 'succeeded' since usually the status quo ante can never be regained. Perhaps a more effective policy is needed?

There are, of course, many studies of the sociology of science and many studies of science and scientists in the policy process which question many of the comfortable assumptions which underlie the more optimistic views of the possibilities of using scientific evidence in a rational manner in discussions about policy. Classic studies of many arenas, notably those of energy policy and choices between energy sources, of climate change and pollution questions and their causes, have shown that much of the optimism about the positive role of science in enabling greater public understanding of complex issues and appreciation of the range and acceptability of different options for solution must be questioned.

Some of the studies have reached to the heart of the objectivity claims underpinning scientists' demands for a voice at the policy table. Sociological studies of how laboratories work, of the conflicts and power plays which govern the very processes of scientific discovery, of scientific fraud and of the ways in which scientific arguments are socially influenced and constructed from a variety of evidential sources, have shown that science is far from the 'neutral' activity that many of its proponents implicitly or explicitly suggest. The most famous example is perhaps the finding of the 'double helix' in the genetic code; that story showed that science as a profession was subject to the play of the same jealousies, pride and unfair games as all other areas, including those of politics itself.

If science is not undertaken using the 'scientific method' – including rigorous analysis of existing data, clear definitions of problems and their situating in broader theoretical systems established through repeatable experimentation leading to public results, tested for accuracy and validity via the system of peer review – much of the legitimacy of those practicing the art of science is called into question. Recent work has also revealed the very partial nature of even that most sacrosanct method of ensuring respectability, the peer review system itself (Price, 1984).

Moreover, it has become clear that most science relating to areas of policy importance is incomplete and unable to provide full explanations for the phenomena perceived, let alone to unambiguously state the social, economic and political importance that governments and public should allocate to the problem in hand. This is especially true of areas where humans have intervened to change natural cycles, natural admixtures of molecules and natural equilibria in the sea, on land or in the air. Of course, since people ceased to be nomads and interrupted the 'natural' ecology from which they gained their food there have been 'unnatural' changes to the ways in which species interact, to which species became dominant, which species became rarer.

Much of what has come to public prominence results from what may be described but not defined as 'excessive' meddling by man with natural processes. Much depends on the acceptance of the idea that there are 'harmless' levels of intervention but that there are also thresholds beyond which harm may come not only to the natural world, defined as distinct from humans, but also to humans themselves.

Deciding what this threshold is and what constitutes 'harm', let alone unacceptable harm, cannot be decided by scientists alone. It is, moreover, an area in which scientists disagree and do so publicly. Their disagreement throws doubt on scientific evidence about the causes of the problem in hand and thus what needs to be done about it. If the experts cannot agree, what hope is there for politicians or the public? Even science-informed policy-making cannot alone find answers to national issues; even less can it provide the answers to problems which demand international cooperation, possibly among a range of actors at very different economic and technological levels. In such circumstances, adopting one solution rather than another will cause differing amounts of social 'harm' and dislocation of established interests, habits and players, even disrupting not only practices but also aspirations for a better life. How galling it must be to developing nations to be told by their richer fellows that the latter recognize that they have created havoc with the natural world on the way to getting rich but that these practices must stop in the interests of the environment of the very same persons and must not be adopted by the persons who seek to emulate the proven roads to wealth.

Detailed studies of the histories of particular areas of policy concern have also shown the ways in which the science concerned is socially constructed. In a very interesting analysis of scientific elites and the making of US policy for climate change research between 1957 and 1974, Hart and Victor (1993) show how the division into disciplines and diverse institutional arrangements where research had different foci contributed to keeping the recognition of damage to the ozone layer and an increase in global warming off the public policy agenda. It also meant that crucial scientific interactions did not occur because some scientists focussed on the oceans whereas others were concentrating on the air with control of the weather on their policy minds. For this and associated reasons it took more than twenty years for the issues to reach public consciousness and for politicians to start taking some of the finding seriously. Thirty years later again the agenda for international action is still in the process of negotiation and at only the very beginning of implementation in most fields.

For the first crucial link between the domains of science itself to occur a number of things had to happen. There had to be a reorientation of the scientists themselves, achieved via the good offices and organizational skills of a well-respected scientist acting as a policy entrepreneur. In addition, the whole arena had to be redefined with 'environmentalism' and 'sustainable development' as the broad heading under which scientists operating in different disciplinary, organizational and funding spheres could come together. Third, wide media coverage had to 'sound the global alarm' which could link the science available to previously rather diffuse public concerns and bring the issues into an arena recognized as one suitable for policy action (Mazur and Lee, 1993).

In their explanation of a chain of events which is similar in many policy spheres, Hart and Victor suggest that any scientific area usually proceeds along its own course, working on problems defined as the most interesting by the senior scientists in the field. There is no guarantee that the streams will converge since the problems defined as the most interesting to scientists may get narrower, not wider, because narrow problems lend themselves better to the central legitimating practices of normal science such as experimentation. Scientific and research careers are based more on increasing certainty than on increasing uncertainty and traditional modes of scientific thought may rule out approaches which go beyond the limits of the dominant paradigm and hence rule inadmissible evidence gathered by other researchers.

Once issues are on the political agenda there are immediately winners and losers from any policy designed to change the status quo. Regulation is often the way in which governments seek to control social behavior in fields where there are potential clashes between public and public interests. Much regulation in complex areas relies on science to define the problem and hence the regulatory solutions which will be effective. This is equally true in what may be defined in the 'positive' regulation areas of health, where new drugs make claims about their effectiveness as disease control mechanisms which need to be tested, and in the more 'negative' areas of pollution control and toxic emissions where activity is more geared to cleaning up messes and preventing further damage, as well as punishing offenders.

The problem for 'regulatory' or 'mandated' science is that its relationship with 'research' science is complex. Levels of proof perfectly acceptable in academic science may be unacceptable to social actors whose livelihoods are affected or to patients suffering adverse side effects from use of a particular drug. Moreover, differences in framing problems to study in the academic and regulatory worlds mean that gaps may appear in the underlying science as some issues were ignored for the purposes of the original experiment or research focus.

Finally here, there are no complete and impartial models of the society to be regulated. This means that policies developed always suffer from the fact that they misjudge or ignore the importance of social processes not immediately obvious to the policy planner who relies on science for his or her direction. Equally, there is usually no single objective which a government is coherently pursuing. Modern governments are diverse and pluralistic and consist of many arms and agencies, often pursuing contradictory policies (Midttun and Baumgartner, 1986). Policy inconsistency is especially common between policies for economic development devised by central agencies such as Treasury, which pursue market-oriented directions, and those aimed at a broader set of concerns, focussing on ensuring citizens' right to clean air, safe water and healthy food. The relative weight to allocate to these activities can only be determined by political processes which move toward compromises which may render scientists furious because their results are being ignored or interpreted in a way which is partial and was never intended.

Studies have also shown that institutional arrangements and national expectations about directions affect the very choice of scientific advice, which is sought and accepted by political decision-makers. Thus, Midttun and Baumgartner show that, faced with the same energy crises in the mid-1970s, different countries at similar levels of industrialization forecast future energy use in quite different ways and made quite different policy prescriptions. Sometimes the attitude depended on the natural resource endowments which countries had at their disposal and which they intended to exploit better in the new circumstances. Accordingly, French forecasts of future energy use were high and encouraged the ever-greater use of energy because the country was ready to develop nuclear power, which it sells on the international market. Much also depended on the political capabilities of the nations concerned.

The French could make the decisions they did because the country had the political and administrative means to enforce that choice on a public which was given few alternatives. In these circumstances, French energy forecasts were virtually self-fulfilling (Midttun and Baumgartner, 1986: 226). In contrast, in the Netherlands, there was much greater public participation in the forecasting process. Community groups, however, had to depend for their data on government sources. In this case, the two sides' forecasts gradually came closer together and the policy developed was a compromise.

Conclusion

The two main parts of this chapter, in their different ways and in complement to the other chapters in this book, illustrate the complexities and uncertainties which governments and the citizenry face in adjusting their policies and practices to an age in which communication technologies of many kinds are increasingly important.

The first part illustrates how governments are using computer technologies in the administrative process itself. It suggests that the reasons for the rapid take up of such technologies are several. On the one hand, information technologies proved the basis for dealing with the enormous increases in the needs of the populations which many governments face. Perhaps even more important, they provide a mechanism for dealing with the enormously enlarged roles which all levels of government play. For most of the period since the end of World War II government in all OECD countries have been expected by their citizens to provide a vast range of services, from the many facets of health to the many aspects of welfare. Governments have to provide these services to their clients in ways which satisfy many criteria, some of which are internally contradictory.

Thus, services should be provided in a timely manner, should encourage the broadest possible uptake by eligible citizens, and should be distributed in an equitable manner. The same citizens also increasingly expect, however, that their services will be delivered in a manner which economizes the taxpayer dollars on which they depend. This means that agencies are expected to deliver the services in the cheapest possible manner. It also means that they are expected to be able to discriminate clearly between people whose claims are legitimate according to the policy in question as interpreted in practice and those whose claims must be rejected. To do this in many cases means that ever more efficient information systems must be developed. We have seen what this development may mean for civil liberties and reduction of inequalities in the society. On the whole, more effort seems to go into finding people who defraud the social security system than who defraud the tax system. This is because public attitudes are themselves often contradictory. Many citizens feel that the tax office is fair game and that the same moral responsibilities do not apply to dubious activity in that sphere. Sensing this, governments seem to place fewer resources in the hands of the tax authorities to pursue fraud than they do in the hands of administrators of social security, notably in the realm of unemployment compensation. Citizens, research has shown, are less willing to tolerate unemployment fraud which they see as 'bludging' on their own hard work.

There is, moreover, no clear distinction within the information systems put in place which could indicate clearly that this piece of equipment and this practice are designed to improve citizen access to their rights or to prevent other citizens from claiming theirs. Much lies in the interpretation of rules and the definition of terms which appear neutral. It is therefore not easy for clients of systems to see the underlying mechanisms which may mean more surveillance of the undeserving or better service for the deserving. Or even the surveillance which serves to distinguish the two. In this sense, too, then governments, just by the use of technologies with great information gathering and storage powers, may narrow the boundaries between public and private information in ways which are not known to citizens.

But citizens may benefit too from the use of the same technologies. We have seen how, as King and Kraemer concluded in 1986, and as I suggested at the beginning of this chapter, computerizing the operations of an agency while it tends to reinforce the prevailing tendencies, leads of itself in no special direction. IT is a tool to command. If the commanders have a particular set of objectives which only pay lip-service to client needs as important to the agency as cost-cutting and preventing fraud, the system they put in place will reflect those values and expectations. If they truly wish to improve clients' access to services or input to policy development, that too is possible.

We may underline this point here by quoting the final pages of Margetts' study. She says that:

> some have predicted that computing will alter the political profile of organisations by shifting power to technocrats .... Others maintain that computing reinforces the status quo by providing the existing power elite with the tools to perpetuate and strengthen their power .... Still others have suggested that computing can strengthen the pluralistic features of organisations by providing different interest groups with the ability to respond to their opposition with the tools of technology. The Operational Strategy appears to provide most evidence for the first two of these. First, by the introduction to the organisation of large numbers of private practice consultants and secondly by the control of acquisition and application of technical skill by the senior management in an already hierarchical organisation. The lack of union or low-level user involvement in the Operational Strategy and the centralised nature of the systems in place have met with little resistance from local office staff armed with technological weapons. Certainly the DSS experience challenges assumptions about the 'democratising' effects of computing. (1991: 338)

In this, says Margetts, her own study merely confirms earlier studies of the introduction of computerization to public sector organizations elsewhere in the world, notably in the US (see Kraemer *et al.*, 1989). Looking historically, Kraemer came to the overwhelming conclusion that the decisions of managers rather than environmental factors have had by far the most significant influence in computing development and use. The case of the DSS shows that the real choice of design lay with senior management and they made the choice with a view of technology as making it possible to carry out a given set of tasks more economically. They did not view it as making possible a whole new set of priorities which may have suggested a new range of policies and a new set of possibilities for fulfilling the needs of claimants, let alone redefining what the needs of claimants might be if considered more broadly.

To some extent, this attitude is part of a wider trend within the public sector. The new managerialism described by Hood has spread throughout most areas of the public service in Australia as it did in Britain in the 1980s (Hood *et al.*, 1990). The emphasis is on the reduction of 'waste'. Internationally there is little call for discussion of the values of administration, which are encapsulated in an emphasis on social justice and access for all eligible claimants as priority values. In a recent paper I have noted the way in which the language of management has taken over in fields for which it was not developed (Marceau, 1996). I also note how that language has taken over not only the implementation side of government but is also critical in defining the agenda for policy. Since I wrote that paper that trend has only intensified. In a similar way, no one mentions the need for justice for social security clients or the need to be more rather than less inclusive in the administration of our programs where policy is made as well as implemented every day. The interpretation of rules is tightened not slackened in cases of doubt. The bureaucratic mode becomes more rather than less formal.

In contrast, however, the same trends may disempower even the bureaucrats. The DSS example shows the extent to which private sector expertise is replacing that of the public administrators themselves. New policies for promotion and recruitment mean that policy expertise is devalued in favor of the generalist managerial skills which suggest that all programs can be designed, funded and implemented by the same modes. If outsourcing goes much further there will be few specific public service functions left. Power will have left the senior levels of all but a few public servants and those who were public officials and leave remain as influential, of not more so, when they are the advisers on industry, taxation or superannuation policies. The ordinary citizen will be served by privatized employment services, private hospitals, the private housing market, private education and private childcare. The wages of the managers will depend on cost-saving and not on service.

To some extent, too, the state may have gone too far in encouraging expression of individual preferences as expressed directly to politicians by means of the new technologies. The research discussed here suggests that removing the broader discussions which can lead to the policy compromises needed for decisions to be made with the overall public interest predominant and favoring of the notion that democracy consists entirely of counting citizen preferences on single issues weakens democracy.

Superficially, this counting process may:

> seem democratic: one gets to make one's opinion known. But the 'one' in this formulation is the privatised viewer rather than the public citizen, and he or she gives an instant 'opinion' rather than concurring with or dissenting from a position hammered out through debate and democratic discourse. A compilation of opinions does not make a civic culture; such a culture demands a deliberative process in which people engage one another as citizens. (Elstain, quoted in Abramson *et al.*, 1988: 176)

Abramson and his colleagues conclude that:

> While political authorities may become more responsive to citizens, especially to mobilised categories, that does not necessarily mean a more responsible politics in terms of the other tasks of leadership in a democracy, such as forging new coalitions out of colliding interests. Nor can one be assured that a greater volume of political information, a large number of political speakers, and the existence of feedback mechanisms will reinforce a near-universal involvement necessary to government by plebiscite. (1988: 163)

Many practical problems are also involved in such experiments. Their solution depends on massive investment by public authorities to increase the extent to which all citizens share equal access to the electronic media and reduce evident inequalities. There is now overwhelming evidence that the world is increasingly divided into the information rich and the information poor. This division runs between societies. It also runs deep within nations, excluding many from the means of participation in the new power system in which information plays an important part.

Here, as noted by Abramson and his colleagues in 1988, we come to an essential irony surrounding the impact of the computer age upon our democracy. In theory, the computer is a vast force for equalizing access to information: no information is so remote or closely held as to be unobtainable by the average citizen. In practice, the computer tends to widen the information gap between economic classes. Much depends on citizens' capacity to subscribe to an Internet service. Public libraries which could have bridged that gap, have been slow to provide such access for reasons of public sector budget cuts. Much also depends on citizens' educated capacity to turn knowledge into information.

In the late 1980s Abramson and his colleagues felt that it was imperative that Congress and local legislatures search for ways to subsidize civic use of videotex and computer information services so as to make the power of the new media generally available to the citizenry. Their call was recently echoed by staff of the Rand Corporation who concluded that all US citizens should be provided with the electronic means of participation in policy-making and its administration at public expense (Anderson *et al.*, 1995).

Governments are also faced now not only with decisions about and administration of complex policies which affect the daily entitlements of every citizen to participate in the reallocation of rewards in a society. They also must deal with increasingly complex issues at a macro-social level.

The issues faced involve the application of knowledge to a much broader range of areas than was the case in earlier decades. At home governments must concern themselves with monitoring and testing the knowledge-based components of their health and welfare systems, such as the drugs which doctors can prescribe or the very training of the doctors themselves. They must also deal with the interactions among the social, economic and technological issues in place and emerging. They often decide which technologies can be used, as in the arena of defense, for example, while in other areas they may determine whether emerging technologies can be used in practice. The use of nuclear power for the generation of energy is but one example. These decisions involve difficult calculations about costs and benefits and the elaboration and implementation of complex systems of monitoring and evaluation of outcomes. They involve understanding risks to public health or safety, devising systems to measure these, and determining which level of risk is 'acceptable'.

For all these tasks governments seek experts, usually defined as scientists who are seen as neutral information holders with no particular axe to grind. In many cases public authorities see themselves as having no choice. However, the issues concerned are rarely purely scientific. Recognizing that immediately leaves the way open for dispute about means, ends and directions. The power of those who control the rules of the game may come to dominate decisions once more.

CHAPTER SIX

Governmental Organization and Implications for Science and Technology Policy

David M. Hart

Governmental organization and S&T policy: an indirect connection

The capacities of a society to create, diffuse, and absorb science and technology (S&T) play an important role in determining whether that society can meet the basic expectations and enhance the standard of living of its members. As recognition of this fact has grown over the past two centuries, and especially over the past fifty years, nation-states have evolved policies to develop and nourish these capacities. The 'geology' of S&T policy since World War II includes 'strata' for military security, improvement of public health, environmental protection, energy independence, and economic prosperity as well as an overlay of concern about education and the advance of culture. Government funding for research and development (R&D) in the OECD countries, which was estimated to be approximately $150 billion in 1998 (OECD, 1999), is the most visible outcropping of national interests in S&T. Trade, economic and social regulation, public procurement, and a host of other policies are among the buried influences on the performance of firms, universities, government laboratories, and other institutions that comprise a society's scientific and technological capacities.

To a large extent, the content of a nation's S&T policies is determined by its level of economic development, its place in the international economy, the degree to which it faces military threats, and other broad features of its society and place in the world. In comparing the S&T policies of the United States and Japan, for instance, one can hardly ignore the United States' status as the world's premier scientific nation, largest economy, and strongest military power. The Japanese rise to technological greatness since World War II was built in part around absorbing American science, serving American consumers, and sharing American security. Such cross-national comparisons are the stock-in-trade of S&T policy studies, and they have proven very enlightening (Ergas, 1987: 191–245; Nelson, 1993).

Yet, such contextual features surely fail to provide a complete explanation of the cross-national policy differences that we observe. Other nations, notably in western Europe, which were comparably positioned to exchange information and to trade with the US as well as to enjoy the fruits of its military spending, did not achieve the same results in S&T policy as Japan. Many factors intercede between the level of economic development and international position on the one hand and S&T policy outcomes on the other. The structure of S&T policy-making institutions is prominent among these factors. The organization of governmental authority, across branches and levels, and the government's relationship to civil society, including the society's scientific and technological expertise and capabilities, has a significant effect on the content of S&T policy. In comparing Japan and western European nations over the past half-century, to continue the example, observers have made much of the eminence of the Japanese bureaucracy, its internal rivalries, and its close relationship with the nation's business leadership. Although the French administrative tradition and the German business–state relationship, to pick two important points of reference, have much in common with those of Japan, no western European nation combines the main features of the Japanese state and state/society relationship. (Freeman, 1988: 330–48; Samuels, 1987; Ziegler, 1997.)

The S&T policy implications of governmental organization are far from direct. The mere fact that a nation has a strong bureaucracy, for instance, does not indicate whether its S&T policies will be more like Japan's or more like the Soviet Union's. None the less, my argument in this chapter supposes that if one could run a social experiment in which one held all other relevant factors besides governmental organization constant, one would find that S&T policies varied systematically. The reason is that S&T are peculiar objects of policy, and its peculiar features interact in different ways with different forms of governmental organization.

One peculiarity is that attention to S&T is at best a necessary, but not sufficient, condition for governments to achieve the major goals for which they are held politically responsible. S&T policy must be integrated into a larger policy for defense, health care, environmental protection or what have you. A related quality is that the benefits and costs of S&T policy tend to emerge only over the long term and that these benefits and costs may well be most closely associated with firms, universities, or other non-governmental institutions through which the policy operates, rather than with the government. These two features of S&T make it difficult for governments to receive credit for good S&T policy and blame for bad S&T policy; particular leaders find it even more difficult to do so. Governmental organization systematically affects the allocation of credit and blame and so systematically influences S&T policy.

Another unusual feature of S&T as objects of policy is that they are inherently unpredictable. S&T policy is therefore a relatively risky approach to problem-solving for governments to emphasize, although this risk can be reduced if governments develop the capacity to learn from the failures that are inevitable for any entity investing in S&T. S&T also change rapidly; policy-makers must hit a moving target to take full advantage of their society's scientific and technological capabilities. These features of S&T require governments to acquire and act expeditiously on scientific and technical information. Governmental organization shapes the way that policy-makers receive and process information, providing another point of systematic interaction with S&T policy.

This chapter explores these ideas in the light of three important forms of institutional variation across polities: the separation of powers among branches of government, pluralism in executive branch decision-making, and the provision of expert advice in the making of S&T policy. I illustrate my point primarily with American examples, reflecting my own knowledge and the biases of the literature, but the concepts and many of the claims are equally applicable in other societies. I close by considering two emerging issues in governmental organization, Federalism and the integration of R&D spending with other policy tools, that have come to the forefront of the agenda for institutional analysis and design in recent years.

The separation of powers: the virtues and vices of complexity

Most of the world's democratic political systems can be categorized as either parliamentary, in which the legislature elects the chief executive, or presidential, in which the chief executive is elected independently. In theory, the parliamentary system provides for more compact and authoritative policy-making, with little dissent, while the presidential system is more cumbersome, leaving open the possibility that the two branches with separate legislative and executive powers may disagree. In practice, there are parliamentary systems that are chronically paralyzed by internal conflict within the government, like Israel (despite the recent changes in electoral law there), and presidential systems in which one branch perennially dominates the other, like France. My concern in this section, however, lies less with the formal structure of government, although the parliamentary/presidential typology provides a useful guide, than with the unity or division of policy-making authority among organizations and individuals with independent sources of power. These organizations may be factions or ministries, as in Japan, or they may be legislative committees and houses, the President, executive agencies, courts of law, and regulatory agencies, as in the US. What matters is that when S&T policy-making power is separated, the policies devised are likely to differ from when it is united. Neither separation nor unity is necessarily preferable in all cases. (Weaver and Rockman, 1993: 1–41.)

In a separated system, for one thing, the power centers compete to take credit for policy successes and to avoid blame for policy failures. Under these circumstances, other things being equal (a condition that is usually implied when it is not made explicit in this chapter, but which is never completely fulfilled), the indirect, long-term contributions of S&T to the achievement of public purposes are likely to be slighted. Policy instruments that promise more immediate and tangible results tend to be favored. In the early days of US environmental policy, for example, the President and Congress jousted to show the public who was tougher on polluters. Instead of providing incentives for industrial innovations that would be likely to significantly reduce effluents and emissions over the long run, policy-makers focussed on requiring firms to meet rigid standards with the best technology available right away with virtually no regard to its cost. The Netherlands' National Environmental Policy Plan, by contrast, relies more heavily on cooperation between industry and government in support of innovation. The unified Dutch governmental system mitigates pressure for immediate results and rigid, inflexible controls. (Hays, 1987: 57–8; Wallace, 1995: 43–61.)

S&T policy may also be a way for separated institutions to appear to take action to address relatively intractable problems, while avoiding criticism for being entirely insensitive to them. The massive US biomedical research program owes some of its success to the need for the political institution most opposed to an interventionist health care policy – sometimes Congress and sometimes the President – to show that it cared about the quality of the people's health. The rejection of President Harry Truman's national health insurance program in the late 1940s, for example, was followed by a surge of Congressional support for the National Institutes of Health (NIH). Similarly, President Ronald Reagan grudgingly supported AIDS research in the 1980s, rather than undertaking the more aggressive public health measures pushed by Democrats in Congress. (Shilts, 1987; Strickland, 1972: 75–6.)

Separated systems are also more likely to engage in distributive politics in S&T than unified systems, particularly when their legislative constituencies are geographically based. The power centers continually vie for the support of regions of the country or sectors of society by offering public benefits or relief from public burdens. For S&T policy, this predilection makes it more difficult to institute the meritocratic approach that the scientific establishment claims to prefer, including such techniques as peer review and delegation of control of S&T programs to expert managers. The National Aeronautics and Space Administration's (NASA) technology development programs, for example, among the biggest S&T items in US Federal budgets, have often been carefully allocated to aid particular firms and regions, especially those with clout in Congress. Although Japanese ministries go to war with one another just as violently as the American military services do, they need to worry much less about the distributional consequences of their S&T programs; Japanese S&T activities are heavily concentrated in the Tokyo area (Cohen and Noll, 1991; Sigurdson and Anderson, 1991: 229–42).

While pleasure is spread around in separated systems, pain is harder to impose on a single place. The US, for instance, has had tremendous difficulty siting a repository for high-level nuclear waste, due in part to the ability of states and localities that might be designated for this dubious honor to appeal through both the executive and legislative branches. The state of Nevada is, at the time of writing, holding off a Congressional majority that voted in favor of placing an interim repository there thanks to the vigorous efforts of its representatives to secure the backing of President Clinton. The more unified systems of France and Japan have had more success in crafting nuclear waste disposal policies, although the problem is far from solved and far from uncontroversial anywhere. (Feigenbaum *et al.*, 1997: 42–109; McCombie, 1997: 56–62.)

The nuclear waste issue demonstrates another proclivity of separated systems in S&T policy: they tend to give more consideration to the uncertainty and complexity inherent in this policy area by stimulating and incorporating diverse views than do unified systems. Members of Congress from Nevada will find and cultivate seismologists, however few or many they might be, who argue that the rock formations where spent nuclear fuel might be emplaced are unlikely to hold up over tens of thousands of years; the US Department of Energy, on the other hand, responding to pressure from the rest of Congress and electric utilities around the country, will tend to uphold an interpretation more charitable to the proposed repository site. The mutual suspicion with which the two branches view each other's experts gave rise in 1972 to the Office of Technology Assessment (OTA), which was intended to strengthen the independent knowledge base upon which Congress could make policy. The abolition of OTA in 1995 does not reflect a diminution of

Congressional suspicion of the President; Congress has many more non-executive sources of expertise now, including staff members, non-governmental advisors, and witnesses at hearings, than it had when OTA was set up. The comparable institution in the French parliament (OPECST), by contrast, is much less-developed than OTA, reflecting the dominance of the executive branch in that system. A stronger OPECST might have provided an authoritative critique of French nuclear testing in recent years; instead, a relatively narrow range of views were considered before the tests were authorized. While unified systems may act more rapidly and forcefully, because there is less dissent, they are prone to making bigger mistakes (as I would characterize the French nuclear testing program) for the same reason. (Bimber, 1996; Holdren, 1992: 235–59; Vig, 1992.)

Decision-making in separated systems tends to be fragmented over time as well as across institutions. No issue ever seems to be settled once and for all. This property of separated systems has a marked effect on S&T policy, because most R&D programs last for several years and have high start-up costs for the people and organizations involved. An uncertain outlook for funding may dissuade researchers from committing themselves to socially desirable lines of work and discourage institutions from establishing linkages that would enhance the utility of the outcomes of this work. In the US, for instance, Congress must appropriate new funds each year, even for multi-year programs (in most cases). As American partners in international 'mega-projects', like the space station and superconducting super-collider, have discovered to their dismay, the money for such projects can be cut off unilaterally, regardless of the President's views or implicit commitments made in negotiations. US participation in such projects is therefore viewed with some nervousness, although the advantages of having the US on board are likely to outweigh such trepidation. The concern about policy instability is not confined to Federally-funded programs; private medical device developers, for example, may have difficulty raising and committing money to R&D if the legal and regulatory environment for the use of their inventions in medical practice is uncertain. Unified systems of S&T policy-making are likely to be more stable over the medium term, since the government can maintain its priorities without challenge. However, when power changes hands in a unified system, complete reversals of course (which are very difficult to bring about in a separated system) are possible. The recently elected Labour government in Great Britain, despite its 'New Labour' label, may well roll back the policies for academic research and other S&T activities of its Conservative predecessors. (Angell, 1996; Dickson, 1997; OECD, 1994: 237–57.)

Both separated and unified systems, then, have advantages and disadvantages in the making of S&T policy. Separated systems are likely to be more deliberate and encompass a broader range of voices and perspectives. They are also likely to be shortsighted, fickle, and prone to capture by particularistic interests. Unified systems, on the other hand, tend to be more narrow-minded and quicker to act, for better or worse, including, once in a while, doing an about-face. But they also tend to have a longer time-horizon, to be more consistent, and to be less permeable to the technically incompetent.

In practice, many S&T policy-making systems have evolved institutions that help them overcome the tendencies to which the larger political systems in which they are embedded incline them, sometimes even overcompensating. Authority over the US nuclear weapons establishment, for instance, was deliberately unified in the late 1940s in the Atomic Energy Commission and the Congressional Joint Committee on Atomic Energy. This cozy arrangement was finally broken down in the mid-1970s by an opposition social movement and an array of perceived environmental and economic mishaps. On the other side of the coin, Japan's Ministry of International Trade and Industry (MITI), in principle an authoritative decision-making body, developed an elaborate system of consultation and 'envisioning' the technological future in order to target government assistance to industrial R&D in its glory days in the 1950s and 1960s. As Japan's high-technology firms caught up their American and European competitors, however, this process was less able to move quickly enough to keep pace with and coordinate private S&T investments. This is not to say that the demands of S&T policy-making drive all national systems toward some golden mean. Rather, my suggestion is that the constitutional structure and institutional traditions that encourage or discourage a separation of powers are important, but far from determinative, influences on the S&T policy process. (Balogh, 1991: 225–94; Callon, 1995; Johnson, 1982).

Pluralism in the executive branch: managed competition or wasteful confusion?

The rather thin discussion in the existing literature about the separation of powers in the making of S&T policy has a more popular cousin in the long-running debate about pluralism in the management of executive branch S&T programs. In some countries, responsibility for S&T is, at least on paper, vested in a single department or ministry with overarching authority. The South Korean Ministry of Science and Technology (MOST), for instance, has such a broad purview. An alternative structure assigns S&T responsibilities primarily to an agency with a particular mission, such as economic development. This structure appears to have been the objective of the reorganization in Great Britain in 1995, in which the Office of Science and Technology was placed in the Department of Trade and Industry. A third model diffuses the executive branch role in S&T among an array of operating agencies with different missions. The US is the cardinal example of such a pluralistic system, in which there is no Department of Science and Technology and six agencies each fund more than $1 billion in R&D. Like the separation of powers, executive branch pluralism and 'departmentalism' have virtues and vices that arise from the unusual features of S&T as objects of policy.

The strongest argument in favor of pluralism takes advantage of the fact that the outcome of any individual science or technology project is unpredictable and its ultimate value is difficult to assess until well after funding has been committed. Progress is not entirely serendipitous, but its pace and direction can rarely be specified in a plan. Pluralism in R&D funding and in technical problem-solving provides for diversification against the risk of failure and increases the odds of unexpected benefits. All three of the American military services, for instance, pursued long-range missile development programs in the 1950s, employing different approaches to this complex technical challenge. The US Air Force, Navy, and Army experimented with various fuels and configurations, before settling on the technologies that became the solid-fuel Minuteman and Polaris intercontinental ballistic missiles (ICBMs). These programs also 'spun off' Atlas, Thor, and Titan, liquid-fuel rockets that proved useful for the civilian space program. (Beard, 1976; McDougall, 1986.)

The pursuit of multiple technical pathways in search of the one that works best is expensive of course, and it creates the potential for bureaucratic conflict. The President and the Secretary of Defense repeatedly reorganized the ICBM programs, appointing numerous missile 'czars', in an effort to save money and resolve turf battles, usually with minimal success. In many cases of technological development, too, the evidence converges at some point on one design that vastly outperforms the alternatives. When such a dominant design can be identified, the most effective strategy is likely to be to concentrate resources on it. Pluralism, however, may make the switch from a multiple path strategy to a single path strategy difficult; vested interests will fight to retain their programs, regardless of the evidence. And, of course, a single big, early bet placed on the right approach is cheapest and quickest of all. The Teller-Ulam design for the hydrogen bomb might be placed in this category, although it is worth noting that the US chose a pluralistic approach even here, setting up two competing laboratories for nuclear weapons design, Los Alamos and Lawrence Livermore National Laboratories. (Rhodes, 1995; Utterback, 1994.)

'Pluralists' and 'departmentalists' also debate whether responsibility for S&T should be integrated with or separated from the larger missions to which S&T are expected to contribute. Research on the process of technological innovation shows that the users of new technologies contribute substantially to innovation and that 'lead' users can fruitfully be engaged quite early in the R&D process. In addition, scientific discoveries may originate from practical challenges as well as from the internal agendas of particular disciplines. These considerations argue for pluralism. Publicly funded agricultural R&D, conducted in the US under the joint auspices of the US Department of Agriculture and stage governments, has produced impressive results in a decentralized framework that engages user participation. Local experiment stations set research agendas appropriate to the climate, soil, and crop conditions in their areas, and extension workers diffuse their results to farmers; it is hard to picture a more centralized system performing more effectively. (Brooks, 1994: 477–86; Evenson, 1982: 233–82; Von Hippel, 1988.)

On the other hand, the inability of advocates of S&T to specify how and when their ideas will prove out may leave them in the cold when an agency needs to show definite progress on a particular problem in a specified time frame, an expectation which is not uncommon (though not always justifiable). Especially when an agency is under political and fiscal pressure, there is a temptation to cut R&D first. US government support for energy-related R&D, for instance, has dropped nearly 80 percent since 1980. Although this decline certainly reflects changes in the energy markets and the end of the Cold War, it also reflects a shortening of the time horizon of the US Department of Energy as it came under external criticism and the weak position of advocates of R&D within the Department. A US Department of S&T might have maintained a portfolio that included more energy-related R&D that did the present pluralistic system. (Holdren, 1997.)

This argument about the relationship of scientific and technological means to governmental ends has a political dimension to it as well. Pluralists argue that a Department of S&T would have a weak political constituency, since it concerns everyone a little, but only the scientific and technical community a lot. The political problems of defending S&T programs within the context of any particular mission, such as energy in the previous example, would be multiplied and the damage of cuts compounded if these programs were consolidated administratively. The affiliation of S&T programs with particular missions may narrow their base of support within the S&T community, but it has the potential to add to the base many non-technical constituencies concerned about the mission. The comparison of NIH and the National Science Foundation (NSF) is instructive in this context. NIH gains when the US public is concerned about health, even though not all scientists care very much about NIH; NSF supports a much broader range of scientific and engineering disciplines, but the public is rarely up in arms about science. NIH, not surprisingly, is about four times the size of NSF. (Intersociety Working Group, 1996: 115.)

'Departmentalists' reply to this argument with the claim that a Department of Science and Technology would give those concerned about S&T policy a stronger voice in setting the national agenda. A seat at the Cabinet table and a visible presence in the budget, in the American context, could create a larger and firmer constituency for S&T. The experience of the World War II-era Office of Scientific Research and Development (OSRD), directed by Vannevar Bush, could be offered in support of this argument. Quickly gaining President Franklin D. Roosevelt's confidence in 1940, Bush had a strong voice in how the nation's scientific and technological institutions and brainpower would be mobilized for war and even inserted himself into strategic discussions that had formerly been the sole province of the uniformed military.[1] (Hart, forthcoming: ch. 5; Reingold, 1987: 299–344.)

In practice, pure departmentalism does not exist. South Korea's MOST, for instance, controls only about a third of that nation's R&D spending; major academic and industrial R&D programs lie outside its influence. The US Department of Defense (DOD) controls more than 50 percent of Federal R&D spending, but it is an amalgam of agencies, including the military services, the Defense Advanced Research Projects Agency (DARPA) and a host of smaller entities.

In order to gain some of the advantages of departmentalism without sacrificing the benefits of pluralism, most governments have developed mechanisms for coordinating, to a greater or lesser degree, the activities of the several agencies implementing S&T policy. The French government, for instance, prepares a national five-year plan for S&T to which agency programs are expected to correspond. Japan has a policy-making Council for Science and Technology (CST) in the Prime Minister's Office that encompasses the main S&T agencies, the Ministry of Finance, and non-governmental representatives of major academic institutions and corporations. The presence of the Ministry of Finance on the CST suggests that, as in most policy areas, budgetary control provides one major tool for governments to coordinate S&T policy. (Baumgarten and Wilsford, 1994; Lederman, 1987: 1125–30; Sigurdson and Anderson, 1991: 49–55.)

The budget process looms particularly large in the US, which has rejected proposals for a more formal S&T policy coordination mechanism, such as a Department of S&T, on many occasions. The Office of Management and Budget (OMB) in the executive Office of the President (EOP) is responsible for evaluating the programs of the Federal operating agencies on an annual basis and nudging them toward greater coherence. The Office of Science and Technology Policy (OSTP, also in the EOP) and various interagency committees that it overseas generally support OMB in this process, occasionally taking the lead role. Of course, Congress does not necessarily respect the President's budget or priorities and can easily disrupt any coordination that has been achieved, if it so chooses. (Dupree; 1987; Hart, 1997; Shapley, 1992.)

As S&T funded by non-governmental sources grows in importance to policy-makers, the task of coordination becomes more difficult. The coordination of private decisions may require legal and regulatory sanctions and incentives that go far beyond R&D funding and engage governmental bodies that have not traditionally taken a role in S&T policy, such as anti-trust enforcement authorities. I will return to this emerging issue below.

Scientific and technical advice: on tap versus on top

The separation of powers and pluralism debates concern the institutional location of S&T policy-making. Scholars and practitioners have also debated extensively the role of experts within S&T policy-making institutions, wherever they might be located in the government. At the heart of the issue is whether experts ought to be viewed primarily as advisors to non-expert political principals, to be 'on tap', so to speak, or whether the experts ought to be 'on top', deferred to by elected officials and appointees. Obviously, the issue becomes moot if experts win election or high-level appointment, but this is uncommon. Even in France, where the political elite is more likely to have scientific or engineering training than in most other countries, the odds of any responsible official having detailed knowledge specific to the technical problem at hand are very low, and expert assistance is typically required. In the US, where scientists and engineers rarely compete for office and are but thinly represented in the top ranks of administration, the need for advice is correspondingly greater. So, too, is the siren song of delegation; American officials have strong incentives to get their knottiest problems off their agenda by handing them over to unelected expert committees. But having experts 'on top' has pitfalls that can be as severe as leaving them 'on tap' and reserving final judgment to those formally accountable to the people.

One dilemma faced in the acquisition of scientific and technical advice is that the interests of experts may differ from those of political authorities. At the crudest level, this divergence may take the form of rent-seeking on the part of experts. Requests for more funding for research, which often accompany pronouncements of advisory committees, may reflect the intrinsic complexity of the problems upon which the experts are pronouncing and the unavoidable failure of prior research to solve them – or it may reflect the desire to keep laboratories afloat and employees in them. President Eisenhower's farewell address in 1960, in which he identified a threat from the 'military-industrial complex' and 'scientific-technological elite', was underlain by the fear that the nuclear arms race could be driven solely by the financial and intellectual appetite of experts and divorced from the political and military interests of the nation. (Crewdson, 1993: 11–16; Sherry, 1995: 233–6.)

A more subtle divergence of expert and non-expert views may arise as a result of the hidden (and perhaps unconscious) value choices embedded in apparently technical problems and the possibility that the values of experts and those of non-experts differ systematically. In the case of AIDS therapy, medical review groups 'on tap' to the Food and Drug Administration (FDA) insisted that FDA fulfill its legal mandate that drugs be shown to be efficacious before being made widely available; the FDA Administrator overruled his advisors in some cases however, agreeing with community activists that offering dying AIDS patients a chance of survival, however small, was more important than scientific certainty. On the other hand, in many instances, expert advisors may legitimately be considered an 'apolitical elite', as Robert C. Wood famously labeled them in 1964, who act on behalf of the public more effectively than high-level officials and should be empowered to do so. NIH's expert 'study sections' translate Congressional support for research on particular diseases into disciplinary agendas on which scientists can act effectively; this research has produced extraordinary advances in molecular biology and biochemistry over the past forty years. Such arrangements can overcome the risk aversion and conservatism that plague more politicized S&T policy-making processes. (Epstein, 1996; Slovic *et al.*, 1982: 141–66; Wood, 1964: 41–72.)

A second pitfall in delegating authority to experts is that they may disagree among themselves. A 'science court' to which public controversies with an overriding S&T content might be assigned, such as that originally proposed by Arthur Kantrowitz in 1967, would at best render split decisions much of the time. Experts may approach the same problem with different levels of confidence in the data and with different models for filling in the gaps. They may even disagree about the nature of the problem and the evidence that would be required to have confidence in a proposed solution. The health risk of ionizing radiation, for instance, is an issue that many polities would like to delegate to export bodies, but for which they typically find that no unambiguous recommendation is forthcoming. Epidemiologists and laboratory biologists weigh the studies from their respective fields differently. In addition, some radiation experts accept that there is a threshold below which radiation is now harmful, while other have adopted the view that there is a linear relationship between radiation dose and biological response regardless of how small the dose is. (Kantrowitz, 1967: 763–4; Mazur, 1974: 243–63.)

When experts disagree, the final policy decision, informed but not resolved by conflicting advice, must be made by administrators or legislators. These non-experts, working with sympathetic experts, typically construct a 'serviceable' version of 'science', as Sheila Jason argues, that stands as the public rationale for their decision. Yet, there may none the less be value in asking experts to attempt to resolve difficult applied questions, even if they disagree at first, because over time a consensus may emerge. If so, this consensus will be even more politically powerful than if it had been present initially. The Intergovernmental Panel on Climate Change (IPCC) seems to be forging a widely accepted position on the environmental consequences of the emissions of certain gases that are produced by common human activities, a near-consensus that did not exist when it started work several years ago. Such a consensus may not put an end to these activities, such as the consumption of fossil fuel, but it should provide a much stronger incentive for governments to move in that direction than they had in the past. In cases like this one, expertise 'on tap' acquires authority that puts strong pressure on those 'on top'. (Intergovernmental Panel on Climate Change, 1995; Jasanoff, 1990.)

Even if experts agree amongst themselves and accurately represent the values of the democracy in which they are operating, there can be miscommunication between those 'on tap' and those 'on top' that disrupts S&T policy-making. Most experts are not accustomed to framing their views in ways that non-experts, particularly political professionals, can relate to their interests and those of their constituencies. Nor are they used to operating in the fishbowl of public and media attention that politicians live with and even crave. The consequence can be confusion about the meaning of findings and controversy where there need not be. Some experts, however, develop keen political skills over time, and the combination of technical credibility and political savvy can be highly effective. President Kennedy's support for the limited nuclear test ban treaty of 1962 for example, was solidified by the advocacy of science advisor Jerome Wiesner, who understood the President's political position, including his campaign statements with regard to the supposed 'missile gap'. Wiesner fused this political understanding with his technical knowledge of test detection capabilities and the prospects for further progress in nuclear weapons technology without testing to build a compelling case that Kennedy should support the treaty, despite opposition from the military. (Brooks, 1964: 73–96; Uyehara, 1966: 112–61.)

Finally, there are certainly cases when the principals 'on top' disagree with their advisors, regardless of the technical evidence 'on tap'. Such cases are painful, either for the expert who must suppress his knowledge or for the principal who must deal with public awareness of internal dissent. In the early 1970s, President Nixon disagreed with some members of the President's Science Advisory Committee (PSAC) about anti-ballistic missile systems and the supersonic transport. The dispute, which some of Nixon's dissenting science advisors made public, probably contributed to PSAC's disbandment in 1973. Nixon chose to do away with the source of pain at the cost of a less-informed S&T policy process. (Smith, 1992: 169–79.)

The rise and fall of PSAC, while poignantly illustrating some of the dilemmas of expert advice, is not representative of the institutional evolution of S&T advisory committees in the US. In fact, they have proliferated steadily, in spite of such difficulties. By Bruce Smith's estimate, some 500 such bodies exist in the executive branch alone; PSAC has even been resurrected, albeit under a different name (the President's Council of Advisors on Science and Technology). As Smith argues, this growth reflects not only the expansion of interest in and dependence on S&T across a wide range of policy domains over the past fifty years, but also the separation of powers and pluralism characteristic of the US. American S&T policy-makers want to make good decisions, but they also need technical support to defend their turf against other branches and other agencies of government and to stake a claim to legitimacy in the public debate. As a result, the US has wrestled with the challenges described above more extensively than have other countries and has tried out a range of devices for managing them, such as the Federal Advisory Committees Act (FACA). FACA was intended by its authors to ensure that advisors' values comport with those of the public by requiring that their deliberations be made public. Critics of the law, which has recently been applied to the National Academy of Sciences for the first time, argue that it has rendered expert advice less useful by breaching the confidential relationship between scientific advisor and political advisee. (Smith, 1992: 1–10; Lawler, 1997: 473.)

Countries with stronger civil service traditions and less fragmented policy-making processes tend to make less use of external S&T advisors and thereby avoid the dilemmas that the use of such experts create. However, these countries are increasingly likely to encounter the sorts of challenges toward which FACA is directed in the US. In France, Great Britain, Japan, and elsewhere, S&T infrastructures are growing, publics are becoming increasingly assertive and less deferential, and S&T-related policy problems remain high on governmental agendas, driving up both the supply of and demand for expert advice.

Federalism: the game goes multilevel

The US, then, is an outliner with respect to the management of expert advisory committees, just as it is with respect to the separation of powers and executive branch pluralism. So it is, too, with respect to federalism, an aspect of governmental organization which has become increasingly salient outside the US in recent years. And, as in the matters discussed above, the American pattern in the case of federalism should not be seen as one that other countries (or groups of countries) are fated to repeat. Instead, the US ought to be seen as an example that might be drawn upon, as either a positive or negative model, for other polities confronting conceptually parallel problems.

These problems grow out of the erosion of the nation-state as the major, if not the sole, level of governance of S&T. S&T policy capabilities have diffused upward to the global and European levels and downward to the sub-national level in the past two decades or so. Perhaps the most important reason for these shifts is the decline in concern about military security, a policy area within which the nation-state was the only legitimate actor, as a driver of S&T policy. At the same time, the emergence of new S&T issue domains, especially environmental and economic policies, should not be underestimated. Trans-boundary environmental issues have created conditions conducive to the establishment of relatively authoritative international regimes for specific aspects of S&T policy. International economic integration is having a similar effect, with the European Union serving as the most dramatic case and the information infrastructure supplying powerful new pressures for coordination, if not unification, of national policies. At the same time, the failure of national macro-economic growth policies and the recognition of distinctive regional industrial competencies in places like Silicon Valley have given birth to a multitude of S&T initiatives at the provincial and state (in the American case) levels. (Haas *et al.* 1993; Saxenian, 1994.)

The resulting multilevel S&T policy game has many features in common with the multi-player S&T policy game in pluralistic systems. When sub-national units are competing against one another, as in economic development, the fact that there are so many of them creates a diverse portfolio of opportunities, brings R&D closer to its users, and broadens the political constituency for S&T policy. At the same time, it creates the danger of wasteful duplication in pursuit of the most promising opportunities; every Alley, Glen, and Hill wants 'Silicon' to precede it. Paradoxically, the presence of more, but smaller, actors also creates the danger of gaps. Provinces and states have weak incentives to support S&T projects that have long time horizons or produce results that are easily appropriable by others. A 'race to the bottom', a familiar phenomenon in federalized policy domains, could easily occur in the support of academic research, for instance, if this responsibility were left to sub-national units.

The dual threats of duplication and gaps inevitably prompt calls for coordination at the national level in competitive S&T policy domains, like economic development. Demands for coordination at the supranational level arise from S&T policy domains that require large-scale resources, such as high-energy physics, or have inherently global consequences, like biodiversity. Supranational mechanisms for S&T policy-making are emerging, albeit much more slowly than their proponents expected when mega-projects and global environmental dangers first hit the policy agenda a quarter of a century ago. The reasons for this sluggish pace are similar to those that stall decision-making in separated systems. The actors, nations in this case, tend to be myopic, narrow-minded and jealous of their sovereignty. The process has many veto points; minority perspectives are not merely voiced, but can typically block action. None the less, when confidence builds at the supranational level or fails at the national level (or both), S&T policy authority may be transferred away from nations. The study of stratosphere ozone depletion and the development and diffusion of mitigation technologies, for example, has proceeded in a relatively coordinated fashion under the Montreal Protocol for the Protection of the Ozone Layer, which was first agreed in 1987 and has since expanded in scope. (Benedick, 1991; Skolnikoff, 1972, 1993.)

Relationships among different levels of governance, between sub-national units and nations or between nations and supranational bodies, vary immensely depending on the policy problem and the historical and institutional context. The federal structure of the US has permitted substantial state-level experimentation with S&T policies for economic development, but the states are only now beginning to develop a means of jointly articulating their interests in Federal policy-making so that gaps can be identified and addressed at the higher level. The shared responsibility of the German Federal and state (Lander) governments in S&T policy, to take another case, has become more explicit with the absorption of the former East Germany. Federal support for industrial R&D is channeled through the 'new Lander', which also take the leading role in higher education and oversee many of the so-called 'Blue List' institutions that conduct applied research and technology transfer activities. The most fully developed supranational S&T policies are those of the European Union (EU). Although the EU funds less than five percent of its members' total R&D activity, its programs have helped begin a cultural transformation of the members' research communities, breaking down barriers not only across national boundaries, but among academic, industrial, and government sectors.

We should be careful not to overestimate the extent to which the nation-state has been displaced in S&T policy. It is still the central arena and will be for a long, long time. Even multinational corporations, which most theorists place at the vanguard of globalization, still display distinctive national traits in their management of S&T, traits that reflect the national policy environment in which they evolved. But there is an evolution in progress, a sharing of and even struggle for power that is making the S&T policy process more complex (Granstrand, 1993: 413–30).

Privatization and policy integration: new tools, new rules

Federalization is just one way in which S&T policy-making is growing more complex. A second set of complications stems from the expansion of the private role in S&T and the recognition that a wide range of policy instruments – well beyond public R&D spending – influences private S&T performance. In the OECD countries, the government share of R&D spending declined from 45 to 36 percent between 1981 and 1995. Much of this decline is attributable to the slowdown in military spending. In addition, the objectives that have replaced national security on the agenda less often require the government to be the end user of the ideas and equipment generated by S&T policy. Indeed, there may be a bias against government end use, as in the pursuit of economic objectives. If firms or other non-governmental organizations are to be the end users, government funding of R&D, if it is provided at all, is only likely to be effective if it is packaged with other incentives for private funding of R&D and for the diffusion of research results and new technologies. The instruments employed in such packages may include tax relief, export promotion, regulatory reform, legal changes, and others. This multifaceted approach to S&T policy brings unfamiliar actors into the policy process and makes the integration of these new actors and instruments with the old ones essential. (OECD, 1996: 290.)

The appropriate S&T policy package will differ from problem to problem, industry to industry, and place to place. An overwhelming body of research suggests that the 'linear model', in which government supports basic research and then lets the free market take its course in converting that research into something useful, is a poor description of reality most of the time. In some areas, like energy efficiency technologies, perverse incentives and large information costs inhibit development and diffusion, even when the technologies are cost-effective. The restructuring of utility regulation and aggressive advertising may be required to overcome these barriers. In other areas, like semiconductors, extraordinary economies of scale and learning can lead to either monopolistic or hyper-competitive market. Monopolies may be stable, but tend to be technologically conservative; hyper-competition may stimulate innovation, but is likely to be extremely wasteful of money and talent. Trade and anti-trust policies might be used to manage such 'strategic' industries in the public interest. In both of these cases, public R&D plays at most a supporting role. And these examples are far from exhaustive. Many other combinations of instruments are possible; rules of thumb for policy design are difficult to come by. (Meyers and Schipper, 1992: 463–505; Mowery, 1983: 27–43; Scherer, 1996: 200–36.)

In attempting to deploy more sophisticated S&T policy packages, traditional S&T policy-makers cross to foreign turf. Economic regulators and anti-trust enforcement authorities, for instance, have not usually weighed their impact on private S&T very heavily in decision-making. They are likely to be more concerned with immediate pricing than with distant technologies. Coordinating their activities with other government programs is difficult; indeed, in the US, regulatory authorities are intentionally de-linked from the executive branch. Yet, without such coordination, S&T policy may be at cross-purposes with itself. The US Department of Energy's R&D program seems ill-prepared to fill the gaps in long-range R&D that are likely to be left by the deregulation of the American electric utility industry.

The set of non-governmental actors concerned with S&T policy also expands as the scope and complexity of S&T policy increase. At a minimum, policy-makers want private sector users of S&T to take an interest in the policy process, but that interest may not be exactly what policy-makers expected. The development of digital television, intended by the US Federal Communications Commission to improve picture quality, may well bring viewers more channels without in the end improving their quality at all. In addition to users, S&T policy may attract the attention of public interest groups and labor unions, not to mention firms and industries that may be indirectly influenced by it. (Brinkley, 1997.)

The need for integration has sparked innovation in S&T policy-making machinery. The recent national information infrastructure (NII) initiative in the US is a case in point The NII and related efforts employed public procurement, deregulation, and intellectual property law along with R&D policy; they also relied heavily on public relations and symbolic action, like the opening of the White House's World Wide Web site. The stakeholders included not merely a dozen or more Federal agencies and Congress, but also states and localities and a broad range of industrial and public interest groups. The center of policy development in this case was a small group within the EOP, including staff from OMB, OSTP, and the National Economic Council (NEC) under the leadership of Vice President Gore. But their reach was greatly extended by the interagency Information Infrastructure Task Force (IITF), chaired by Secretary of Commerce Ron Brown, which engaged the business community in particular. Although IITF was not fully successful and should not be taken as a well worked-out model, it may indicate a future direction for S&T policy processes: more open, looser, and more flexible and adaptive. Such arrangements are probably easier to put in place in the American political system than in others. (Hart, 1997; Kahin, 1997.)

As with federalization, it is important not to exaggerate the extent to which new rules and new tools are supplanting what some critics have called the '1950 model' or the 'old paradigm' of S&T policy. Government R&D spending, for better or worse, still dominates the debate, the military R&D still comprises the biggest share of it, in France and Great Britain as well as in the US. The old S&T policy paradigm, like any old paradigm, will not disappear without a fight. In fact, it may not be accurate to think about the process of institutional change in this sphere as a paradigm shift at all. A more satisfying metaphor is that of tectonic movement, in which the various 'strata' of S&T policy, to which I referred at the outset of this chapter, are becoming independent landforms on the policy map (Crow, 1994).

Conclusion: the challenges of institutional design

The main features of governmental organization for S&T policy are not matters of choice for S&T policy-makers. Federalism, the size and quality of the civil service, executive branch pluralism, the separation of powers: these stem from national constitutions, cultural traditions, and cataclysmic historical experiences of war and upheaval that transcend the specifics of any policy area. Their implications for S&T policy are to a certain degree unavoidable, and those involved in any national policy process must simply make the best of their organizational endowment.

Yet, as this chapter has noted, many national S&T policy processes exhibit signs of evolution that suggest that governmental organizations are not completely rigid. Pluralistic systems have developed coordinating capabilities, while centralized systems have shown some capacity for flexibility. The worst implications of governmental organization for S&T policy may thus be avoided or, perhaps, once they are experienced, policy-makers make adjustments so that bad experiences are not repeated. One may hope that these evolutionary changes can be guided more consciously by analysis and learning than they have been to date. To the extent that I have identified important sources of systematic variation in S&T policy-making, such as the ways that political systems allocate credit and blame and acquire and process information, this chapter may serve as a starting point for institutional designers.

Institutional design is a difficult and often thankless task. Vested interests inevitably obstruct the implementation of new designs. The results or restructuring are difficult to assess and easy to distort. At best, the influence of institutional designers is felt indirectly and over the course of many years. Yet, these difficulties make their work, and our work as institutional analysts, all the more vital. Systematic changes in governmental organization, which redistribute credit and blame or change the gathering and management of information, can be deep geologic forces that shape the contours not merely of a single obscure policy subsystem, but of defense, health, environmental, energy, economic, information, and whatever other major policy areas the future adds to the world of S&T policy.

Notes

The author thanks Lewis Branscomb, Harvey Brooks, and David Guston for their comments on earlier drafts.
1. It should be noted that Bush himself was a strong pluralist and believed that the war was an extraordinary situation that justified extraordinary administrative arrangements.

CHAPTER SEVEN

Autonomy and Accountability for Twenty-first-century Science[1]

Susan E. Cozzens

Introduction

Among the most puzzling paradoxes of the knowledge society is the simultaneous need for more autonomy *and* more accountability in science. In relation to science, the word autonomy brings forth an image of freedom from external control, the provision of resources with 'no strings attached'. The word accountability, in contrast, brings to mind those very strings: paperwork, audits, and the idea of delivering on a contract. In this chapter, I argue that beyond the surface these two seemingly contradictory concepts are in fact compatible. Furthermore, the ways of combining them that have evolved in research policy may serve as examples for governance in the twenty-first century in other areas. In the next section, I explain why the knowledge society needs more of both autonomy and accountability. I then review the history between the two, as mediated through government funding for science in the United States, and describe in what sense each needs to be deeper in the knowledge society. In the conclusion, I indicate how lessons learned in research policy about decentralized democratic governance might be applied in other areas.

Why both?

The need for autonomy for science in the knowledge society is often justified in economic terms. Autonomy is seen as the source of creativity and innovation; scientists need to be left free to explore whatever leads they want if they are going to think new things. And the knowledge-based economy needs new thoughts. A knowledge-based economy depends on breakthrough products and technologies. If it merely improves incrementally on what it is already producing, it will rapidly lose competitive edge, as new product areas are opened up. So it needs innovation and creativity on the part of its knowledge-procedures just to stay alive.

Financial considerations also explain why contemporary societies need accountability. In the societies that are becoming knowledge-based, government pays for much of science. And in all these countries, governments have limited budgets to spend. Those budgets are squeezed on one side by the entitlements promised under the welfare state in its various forms, and on the other, in some countries including the United States, by military expenditures. These dual pressures have decreased the percentage of public spending available to 'discretionary' activities like research; in the United States, that share is now under twenty percent. Under these circumstances, each item that competes for 'discretionary' funding must be able to demonstrate its benefits to the public. Thus the surface pressures for accountability are increased. In some countries they have been appearing in the form of legislation that demands performance indicators and other forms of reporting.

There are deeper reasons, however, for knowledge societies to need both autonomy and accountability. We are referring, after all, to a knowledge society, not just a knowledge-based economy. For a society as a whole to be truly knowledge-based, creativity and innovation must be shared across the society, not just the province of a small, privileged group. Researchers cannot create a knowledge society on their own. Rather, everyone in that society must be knowledge-activated, creativity-activated, for the society and economy to be dynamic. I call this state of affairs shared creativity, and I will return to it at the end of this chapter.

Autonomy for prosperity

The condition of shared creativity does not come naturally in the institutional structures within which science now lives. Those institutions were set up with quite a different set of assumptions in mind than the ones just articulated. They were set up under what I will call the autonomy-for-prosperity model of public funding for scientists. In the United States, we trace this model to Vannevar Bush, and his efforts to set up a National Research Foundation in the late 1940s. Let me review that model briefly.

As long as there have been discussions around science policy, the fundamental rationale for government support for science has been the achievement of public goals. In the United States, Europe, and Australia, for example, government was lured into increasing support for science by the overwhelming contributions of scientists to the victory in World War II. Developing countries have likewise often sought to stimulate their economies through building indigenous research capacity. Although nations have varied in whether they leave goals general or put them in specific terms, and although national goals have varied over time and across countries, the overall objective of putting knowledge in service to society has been universal.

The funding mechanisms and program forms that have been invented to accomplish that end, however, are far from universal. In Europe, funding strategy has been linked to the presence of national university systems, and the growth of the European Union has added a further level of complexity along with a new set of resources and constraints. In France and in the socialist countries, centrally administered government laboratories took center stage. The United States, in contrast, has experimented over time with a mix of program support in government laboratories and project support for university research, with the laboratories generally more focussed on applied research and technology projects and the universities focussed on fundamental research.

Vannevar Bush proposed the second element of this mix, project support for universities, in the late 1940s in a document called *Science, the Endless Frontier*. Bush had been a leader in the effort to use scientists to help win the war, and he understood quite well the programmatic, targeted approach to applying knowledge in service to society. In the Manhattan Project, which developed the atomic bomb, government set a specific objective, and scientists responded to solve the problem, both using what they already knew and creating new knowledge. But after the war, Bush sought quite a different relationship between the research community and government. He argued that research needed a freer form of support, in order to build the kind of capacity that was put to work in the Manhattan project and to explore the new ideas that would produce future advances. The knowledge and human resource base for meeting national security, health, and economic needs in the postwar period, he maintained, should be expanded through government funding for non-targeted, non-programmatic science.

In *Science, the Endless Frontier,* Bush put forward the autonomy-for-prosperity model of the relationship between research and the public. Bush's view of prosperity was far-reaching:

> Advances in science when put to practical use mean more jobs, higher wages, shorter hours, more abundant crops, more leisure for recreation, for study, for learning how to live without the deadening drudgery which has been the burden of the common man for ages past. Advances in science will also bring higher standards of living, will lead to the prevention or cure of diseases, will promote conservation of our limited national resources, and will assure means of defense against aggression. (Bush, 1945: 10)

By autonomy, he meant freedom from targeted programs:

> Many of the lessons learned in the wartime application of science under government can be profitably applied in peace. The government is peculiarly fitted to perform certain functions, such as the coordination and support of broad programs on problems of great national importance. But we must proceed with caution in carrying over the methods, which work in wartime, to the very different conditions of peace. We must remove the rigid controls which we have had to impose, and recover freedom of inquiry and that healthy competitive scientific spirit so necessary for expansion of the frontiers of scientific knowledge.
>
> Scientific progress on a broad front results from the free play of free intellects, working on subjects of their own choice, in the manner dictated by their curiosity for exploration of the unknown. Freedom of inquiry must be preserved under any plan for government support of science .... As long as [colleges, universities, and research institutes] are vigorous and healthy and their scientists are free to pursue the truth wherever it may lead, there will be a flow of new scientific knowledge to those who can apply it to practical problems in government, in industry, or elsewhere. (Bush, 1945: 12)

Most of the experimentation over the next few decades with ways of placing science in service to society took the form of add-ons to this basic model of autonomy-for-prosperity.

The autonomy-for-prosperity model drew on an earlier model of autonomy-for-truth embodied in the university; but there are profound differences. The autonomy-for-truth model, articulated in the German university in the nineteenth century, and elaborated by sociologists of science in the 1950s (Barber, 1962), maintains that research should have autonomy in order to be able to achieve objectivity. To have a truth-producing sector of national life, a country must allow that freedom. The autonomy-for-prosperity model, in contrast, says that research should have autonomy because this is the best route to an improved quality of life. Vannevar Bush was a pragmatist, not an epistemologist; his primary concern was invention, not truth.

Bush's organizational designs, developed during this period, built a shell of organizational autonomy around research to protect it from the level of political influence that scientists experienced during the war and that many feared would come with further government funding. To understand the meaning of this protective organizational layer, we need to pause briefly to distinguish two types of autonomy for research. The core meaning of the term is at the level of individuals or teams (Cozzens, 1990). If the economy requires new thinking, then autonomy must be provided to thinkers. However, autonomous thinkers have always needed some form of institutional protection. Before governments got involved in research, the university provided that protective institutional shell. No existing government agency seemed likely to provide this level of insulation for research, however, so Bush invented a new one. The second meaning of the term autonomy refers to the operation of the organizations that provide the protective shell.

Organized along the lines of the Republic of Science, Bush's new design called for scientists themselves to control government funding organizations. The National Research Foundation that he proposed was to take a long-range view, not confine itself to short-term priorities. It was to have non-political leadership and rely entirely on external organizations to accomplish its goals, never build up laboratories of its own. Furthermore, it was to leave control of the research agenda to the research-performing institutions, largely universities. In short, it was a nurturing, not a leading organization, as befit its purpose of building capacity and allowing new ideas to emerge.

One of the major struggles over the establishment of this organization became its low level of political accountability. Bush's original design called for a self-perpetuating board, which would none the less be responsible to the President and Congress through normal audits, reports, and the budgeting process. The President took exception to this plan, and insisted on making appointments to the Board himself (England, 1983). While the plan for the National Science Foundation was held up in political debate, plans for the Office of Naval Research and National Institutes of Health, which embodied many of the same principles, with variations, moved forward less problematically (Sapolsky, 1990; Strickland, 1989).

The autonomy-protecting organizations set up during this period have proved remarkably resilient. NIH in particular has managed to balance autonomy and accountability in some remarkable ways, skirmishing over the macro-shaping of its budget while welcoming lay members into its funding councils and maintaining the scientific purity of the peer review process. It has generated enough political support to be the growth leader among all government research agencies in the past five decades, yet has maintained an extraordinary level of internal control.

Two funding mechanisms embody the essence of the autonomy-for-prosperity model. One is the project grant/peer review system of funding. By accepting project proposals, the government hands over problem formulation to researchers themselves, leaving a significant space for undirected creativity. By selecting from among those projects primarily on the advice of peer reviewers, government also leaves the judgment of what is to count as creative in the hands of the research community. The second funding mechanism, science development programs, is even freer. Set up in the late 1950s, these programs distributed large blocks of funds to be used completely at the discretion of universities to build their educational and research capacities. By pouring millions of dollars into universities through the 1960s, these programs greatly expanded the number of campuses capable of supporting first-class research.

When the autonomy-for-prosperity model came under attack in the 1960s (more below), a visual representation of it appeared that has survived and thrived. This is the spider-web diagram, which traces the 'critical events' that led to some recent technological advance back into earlier technologies and research efforts. This art form was born of conflict and controversy. In the mid-1960s, a Defense Department sponsored study (Hindsight) traced the origins of defense technologies with this technique, and found that technology grows largely out of earlier technology. NSF answered with TRACES, a study that extended the search for critical events further back in time, and not surprisingly found origins in fundamental research. A similar study in biomedicine produced similar findings on the crucial role of non-targeted research. Such diagrams are still considered the most powerful communicators of a basic message: We must build a generic base of knowledge and capability, or we will not be able to innovate.

But this powerful visual representation of the claim that the autonomy-for-prosperity model also illustrates its mystery. Why does this model work? It specifies no mechanisms for putting knowledge into service to society. For example, how do problem-solvers learn about the research knowledge that is available? Some analysts have claimed that the reason the autonomy-for-prosperity model worked in the 1950s is that scientists at that time were steeped in the practical problems of the war period: radar, medicines, etc. They were still doing what one industrial research leader calls 'grounded pioneering research' (Brown, 1995). One can argue that NIH's success springs from a similar source: many of its investigators are also practicing physicians, or at least located in medical centers. They are thus thoroughly immersed in the practical problems of medicine, and can choose and interpret their research projects effectively in the light of medical needs.

Ironically, the contact with practical problems that produces 'grounded pioneering research' was undermined by the very effectiveness of the autonomy-protecting institutional shell Bush designed. As government funding for research grew rapidly in the 1950s, beyond the scale of Vannevar Bush's wildest dreams, the institutional shell inadvertently created a protected space where new researchers could exist without any grounding in the practical problems of the world. As soon as the practical grounding was lost, a gap was created between science and society that needed to be bridged if autonomy was really going to be turned into prosperity.

Knowledge transfer

Among the first set of methods developed to bridge this gap was a family I will refer to as 'knowledge transfer' mechanisms. Knowledge transfer mechanisms do not threaten either individual/team-level or organizational autonomy, because they leave the protective shell in place while focussing on diffusing or disseminating research-based knowledge.

The national interest in the problem of knowledge transfer was recognized early, soon after the federally funded expansion of the knowledge-production effort began. In 1958, for example, a high-level group recommended the formation of an Office of

Science Information Service in the National Science Foundation (Hillman, 1982: 61). In the early 1960s, several influential voices argued that government should take responsibility for disseminating research results and maintaining an infrastructure that would allow their application. In the late 1960s, a branch of the National Academy of Sciences recommended that private organizations be supported by the government effort, including scientific and technical societies (Hillman, 1982: 61).

Steps toward sharing of research knowledge became part of the response to social questioning of science in the 1960s. Communication was definitely one of the issues. A Presidential science advisor observed

> now many Congressmen and members of the public ... see a scientific community which, insisting on its purity, will not deign to communicate with the public and justify itself, but prefers to believe that its virtues are so self-evident that a right-minded society must necessarily support it on its own terms. (Hornig, 1968: 248)

An OECD task force on information policy, reporting in 1971, likewise saw the public as a crucial audience to be reached by information policy and stressed the use of scientific and technical information in public debate.

> Just as science policy must be an integral part of overall government policy, linked to social and economic affairs, information policy must seek to assure that the world's specialized and professional knowledge is fully and properly used in guiding social evolution. ... The ability of society to use technology wisely is at stake (OECD, 1971a: 18).

Throughout the history of information policy, however, the emphasis has been on providing infrastructure for communication, not shaping its content. Science was to speak to the public, not with it. First government encouraged journals; then abstracting and indexing services, to provide access to the exploding journal literature (Bortnick and Miller, 1986: 354). Extension services were expanded later: the 1965 State Technical Services Act, for example, stressed exchange of technical information among industry, universities, and state governments (Smith, 1990: 88). In the early 1980s, NSF's efforts focussed on design of information systems, the algorithms and tools that could be used for retrieval. Eventually, the federal investment evolved into support for an Information Superhighway. In each case, these policy efforts put an infrastructure in place to distribute research-based knowledge; and in each case, the emphasis was on the pathway, not on the knowledge generation process itself. Knowledge transfer mechanisms thus leave both forms of autonomy intact by segmenting knowledge processes sequentially, in time.

Researchers still often voice this strategy as sufficient for accountability. The full responsibility of researchers, in this view, is to make their findings available in the public sphere through publication. It is then the job of society to use the knowledge. If this view of accountability were sufficient, it would greatly simplify the job of developing performance indicators for research organizations: publications alone would be enough. Unfortunately, that indicator will probably not satisfy the most important audiences for accountability legislation. Congressional staff members have observed, 'We don't know exactly what we want, but publications in peer-reviewed journals will certainly not be enough.' 'Where's the economic growth?' another set of staff members asked in discussions around research performance plans based on peer review. State S&T programs collect publication information, but find that job creation is the primary indicator state legislators want to see (Cozzent and Melkers, 1998).

Knowledge mandating

As the size of the research enterprise continued to grow through the 1960s, the combination of peer-reviewed project support and block funding to universities eventually generated a backlash. In 1969, Congress, clearly judging that the autonomy-protecting model had spread too far, pasted a rider on an appropriations bill restricting the Defense Department to supporting goal-directed research. Called the 'Mansfield Amendment', this measure sent shock waves through the newly expanded basic research community. Many leaders of that community saw the separation between goal-directed and basic research as artificial. The National Science Board, for example, passed a resolution calling for agencies to continue to support 'their' basic research (Smith, 1990: 81–2). But outrage among scientists did not carry the day. The backlash reached its crescendo during the early 1970s, when the Nixon administration phased out science development programs (Smith, 1990: 82).

At the same time, a new wave of targeted programs appeared. These put in place a process which, following Machlup (1980), I will call knowledge mandating. Knowledge mandating is the active calling-forth of knowledge, rather than passive waiting for the right knowledge to appear. Two examples of US targeted programs during this period, the War on Cancer at NIH and the NSF program of Research Applied to National Needs (RANN), are particularly well known.

The RANN program grew out of a new dialogue between the President, Congress, and the National Science Foundation. Lyndon Johnson is reported to have expressed concern over 'too much research being done for the sake of research' (Mogee, 1973: 28). Under Nixon, pressures on the Foundation to incorporate problem-oriented research came from both administration and Congress. NSF's first programmatic response was under the title, 'Interdisciplinary Research Relevant to Problems of Our Society' (IRRPOS). The word *relevant* was carefully chosen by the NSF Director, who feared that if the research were *on* societal problems, 'the new program would soon be expected to produce tangible results' (Mogee, 1973: 42). NSF's first efforts in this direction faced 'a wall of hostility' (Mogee: 50) from the rest of the Foundation's staff, but Congress considered them too weak. In the next budget round, the administration tied a big increase in funding for basic research to establishment of a new program with an 'active, aggressive' (Mogee: 66) management approach, specifically defining and delineating societal problems for attack and following project results closely. NSF quickly accepted the deal, and RANN was born. At the recommendation of the National Academy of Engineering, this 'top-down' approach was segregated into a new organizational unit.

The War on Cancer emerged during the same period. President Nixon's State of the Union address in 1971 included 'an appropriation of an extra $100 million to launch an intensive campaign to find a cure for cancer' (US Senate, 1971: 74). In this, Nixon was responding to clear indications that Congress would establish the program on its own if he did not propose it first (Rettig, 1977). The relevant cabinet member warned that 'dollars and large-scale organization are [not] an adequate substitute for ideas and a sound scientific base', and NIH Director Shannon expressed concern that skewing money toward a particular area would retard the development of science (Studer and Chubin, 1980: 77–8). But Congress wanted results. As a first step, it established the Special Virus Leukemia Program, which 'engaged in more contract research than had been the custom of any NIH agency. . . . Such contract research reflected a new type of planning premised on the operations research strategies that had worked for other areas of science' (Studer and Chubin: 78).

RANN and the War on Cancer are the best-known targeted programs of this era, but there were many other related developments at the time. Where contracts, problem delineation, and strong project management were involved, the individual/ team autonomy of researchers was clearly threatened by this set of developments. But sometimes, softer programming simply shifted resources from one broad area to another. In soft programs, individual/team autonomy remained intact, but organizational autonomy was threatened, since in this instance, external pressures were dictating to the supposedly scientist-led organizations how they should allocate their resources.

The history of these efforts in the United States carries an important lesson. Over time, organizational autonomy won out over societal knowledge mandating. RANN, for example, was eventually abolished, and its funds were reabsorbed into the base level of funding for the National Science Foundation. Over the years, both NSF and NIH have found ways to look programmatic without doing central planning. Indeed, while it seems on the surface that programmatic mechanisms continue to be used in the United States, in reality they are getting weaker. A case in point is the elaborate interagency coordinating mechanisms that appeared in the late Bush administration and were adopted widely in the early years of the Clinton administration, including the High Performance Computing and Communications initiative (HPCC) and the US Global Change Research Program (USCGRP). On the ground, there has been little enthusiasm for these mechanisms within agencies unless they hold the promise of increased funding. And so far there have been no serious attempts to check on the results for society of these 'programs'. I take both these observations as signs that the mechanisms actually exist for science, not for society.

The society-oriented program structures of US research policy, in the end, has not turned out to be a set of fixed channels that actually direct resources toward national priorities. Rather, they are a kind of transparent overlay which, placed on top of existing activities, changes the interpretation of the pattern and shows how fundamental research serves society. The overlay strategy is completely legitimate and reveals a reality that would be unnecessarily hidden without it. The point is, however, that this kind of programmatic structure does not constitute knowledge mandating; it actually reveals that the autonomy-for-prosperity model is at work.

Knowledge sharing

By the mid-1970s, a third approach began to be developed in the United States to bridge the gap between science produced inside the protective institutional shell and the problems of the world surrounding it. This new form threatened neither the individual/team autonomy of scientists nor the organizational autonomy of funding organizations. I will refer to it as knowledge sharing.

This approach first took the form of an emphasis on partnerships, and within that emphasis, an early focus on partnerships with industry. In the US, this era was heralded by the appearance at NSF of Industry-University Cooperative Research Centers, small centers on campus that created sites and incentives for interaction. These were soon followed by the much larger Engineering Research Centers, which spent millions annually. The ERC program had an ambitious goal: to change the culture of engineering research and education in the US. ERCs involved industry partners in setting strategic plans, and interchange of people and information became the rule. The critical element was a two-way dialogue, which replaced the one-way science-to-society diffusion of information that characterized the knowledge transfer mechanisms and the one-way society-to-science mechanism of knowledge mandating. In the two-way dialogue scientists became strategic thinkers, able to formulate their own problems creatively, but steeped again, as in the 1950s, in the problems of some external set of partners. Another new element was the link to education, at both graduate and undergraduate levels. The ERCs were intended to produce a 'new breed' of engineers, better prepared than previous generations to participate in R&D in industry because they understood the needs and culture of industry (Parker, 1997).

In the late 1980s, a second crucial step in the development of the partnership model took place with NSF's establishment of Science and Technology Centers (STCs). STCs were urged to form partnerships not only with industry, but also with state and local governments, citizen groups, and schools. The benefits that industry gained from ERCs were now available to other parts of society. Strategic plans began to be shaped by two-way dialogue across many sectors, and the education of a new breed of researcher able to bridge the cultures of university, school, and public service sector began. After ten years of success, the STC program has been approved for another round by NSF's governing body, and encourages institutions to reach even further and prepare students for an even broader set of contacts and careers. As NSF describes the new program,

> Science and Technology Centers build intellectual and physical infrastructure within and between disciplines, weaving together knowledge creation, knowledge integration, and knowledge transfer.... New knowledge thus created is meaningfully linked to society. (NSF, 1997: 2)

The STCs represent a phenomenon I will call partner pluralism. I will return in the next section to the crucial role of this pattern in the knowledge society.

A second manifestation of knowledge sharing has arrived with the recent round of attention among researchers to public awareness of science and science education. It is easy to be cynical about any effort in science education on the part of the research community. Greenberg has described these activities as purely public relations efforts on the part of research. There is, none the less, a new note in the recent discussions on this topic among science leaders. Perhaps the new note is one of desperation: there has never been a time when science needs the public more than it does now (Greenwood, 1996).

Whatever the reason, the fact it that research leaders on more than one occasion recently have stressed that scientists need to learn more about the public, as well as the public learning more about science. For instance, Mary Woolley of the science lobby group Research! America writes:

> There are many ways for a scientist to become involved in the local community – through visits to schools or guest lectures at civic clubs, churches, and senior centers, for example. Ultimately, every scientist should be able to answer 'yes' to what I call the '7– 11 test': Do your community leaders recognize you by sight and know your issues when you see each other in the local convenience store? (Woolley, 1997: 53)

Even the venerable National Academy of Sciences is recommending that the institutions of science open their doors to the public, with recent urging from the Institute of Medicine that NIH take more advice from the public in priority-setting. These are the signs of the beginning of a two-way dialogue, and lay the groundwork for the full development of creativity sharing.

Creativity sharing

I began this chapter by claiming that the knowledge society needed to achieve a state that I called creativity sharing, which we can now understand as a stage of development of knowledge-sharing. Let me give a fuller definition of this critical concept:

> Creativity sharing is a state of dynamic, collaborative relationships between scientists and a wide range of other social actors that facilitates joint definition of problems that need research solutions and joint development of those solutions.

I do not know exactly what this coming state of creativity sharing will look like, but after the preceding analysis, we can see some of the elements of the current situation that form the seeds of this later stage, and from which it is likely to grow. The analysis also leads us to identity the next steps that need to be taken in the direction of creativity sharing.

One seed for creativity sharing is partner pluralism; that is, the extension of partnerships for researchers beyond academe and beyond industry to other societal actors. The importance of this extension can scarcely be overstated. Individual/team autonomy is not threatened by exclusively industrial partnerships; there is every sign that individuals and teams can be very creative within the broad problems set by commercial development, and can continue to think the new thoughts that can be developed into new products and services. There is also ample reason to believe, however, that new products and services developed in this way will only haphazardly, if at all, lead to stronger communities with greater quality of life for all.

To keep these larger considerations alive in the innovation process, public research funding organizations have critical roles to play. These roles include envisioning and exploring technical options regardless of short-term profitability and maintaining the public data and knowledge bases that underpin that exploration. But the public orientation of these organizations is threatened by exclusively industrial partnerships. Government research funding organizations need partner pluralism to maintain their public character and avoid becoming mere extensions of private

R&D laboratories. The next step toward shared creativity is thus to extend partner pluralism to the governance of such organizations.

A second seed from which shared creativity is growing is science education, both formal and informal. The goal of this effort must not be one-way diffusion of knowledge, nor the memorization by the public of certain scientific 'facts'. This concept of science education as tested by the pop-quiz science literacy surveys does not provide any base for creativity sharing. Instead, the goal of science education for the knowledge society, must be activation of all citizens. That is, we must give each citizen a base sense of competence in science-based issues that removes any inhibitions to participating in knowledge partnerships and indeed turns everyone into an active problem-identifier and problem-solver. The crucial partners for universities in achieving this goal of universal activation are schools. Indeed an excellent next step toward the knowledge society along this route, along with a model of creativity sharing, would be for universities to engage teachers and students in schools in the process of developing educational technology, collaboratively, rather than as a one-way technology push.

The third seed from which creativity sharing is growing is the civic scientist, a concept introduced and championed by Neal Lane, past director of the National Science Foundation and then Science Advisor to President Clinton. The civic scientist is prepared for a flexible career inside and outside academe, has an attitude of service toward society and embraces teaching and public outreach as part of his or her vocation. As Lane (1998: 14) writes:

> In this new civic capacity, scientists and engineers step beyond their campuses, laboratories, and institutes and into the center of their communities to engage in active dialogue with their fellow citizens. ... Do I mean that we go out and teach science to shopkeepers, lawyers, consultants, and construction workers? Not entirely. To engage in dialogue is to listen as well at to speak. While there is a great need for the public to have a better understanding of science ... there is as great a need for scientists to have a better understanding of the public.

The next step in fostering civic scientists for the twenty-first century is the reform of graduate education, a large project which many universities have already begun and a trend which government agencies are now attempting to strengthen.

Shared governance in the knowledge society

In conclusion, then, we see that accountability in the knowledge society may take the form of partnerships, the engagement of many societal actors in the autonomous exercise of shared creativity. The keys to shared creativity are equal footing and two-way dialogue. For society to achieve this state, scientists must become strategic thinkers steeped in society's problems and issues, and government funding agencies must remain public through partner pluralism. Without these characteristics, we will be left with only a knowledge-based economy, not a true knowledge society. Knowledge-based economies may maintain their innovation rates, but at the same time they will experience internal tensions and rigidities that will lead them, over time, into temporary economic successes that undermine the long-term development of sustainable economies. In contrast, true knowledge societies will be effective because they are more broadly innovative, and will develop new products and services that serve the common good. These more flexible and inclusive societies are likely to be the winners in the global economy over the next few decades.

The decentralized, democratic character of shared creativity may make it a model for other areas of governance in the twenty-first century. If we agree that decisions about technology are crucially shaping our futures, then much is at stake for the public in choices made within private firms. National governments and international policies cannot keep pace with the changes, and the traditional regulatory frameworks that assure safety and effectiveness develop at a much slower pace than the technologies they address. Central government is thus losing some of its effectiveness as a forum for public choice, at least in the technological arena. We need a way to incorporate public interests into private action without centralized government intervention. Local dialogues of the sort developing in research may be just the kind of decentralized democracy the twenty-first century needs.

This thought is reflected in a vision Vice-President Al Gore articulated in 1997. He compared the transformation of knowledge today to the shift from mainframe to parallel processing. In parallel processing,

> when a problem was presented, all of the processors would begin working simultaneously, each performing its small part of the task and sending its portion of the answer to be collated with the rest of the work that was going on. It turns out that this 'distributed intelligence' approach is more effective for solving most problems. (Gore, 1997)

Why has this concept never spread to social problem-solving and politics, he asks. It not only helps explain why democracy has triumphed over government that rely exclusively on a central authority, but it helps explain why corporations that transfer authority to front line workers are becoming more effective.

Here at the edge of a new century, we have a choice of two paths. One path retreats from understanding, flinches in the face of challenges, and disdains learning. It leads to a know-nothing society in which the storehouses of knowledge dwindle, the spigots of discovery are turned off, and missions of exploration are stalled on the ground. ... But there's another path – infinitely brighter and considerably more American. It leads to a learning society whose government continues to fund basic research and applied technology and in which the virtuous circle of progress and prosperity is alive and functioning (Gore, 1997).

It is within our power, he said, to blaze the second trail. Scientists and engineers, as the very embodiment of that ideal, have an obligation to make it happen.

Note

1. An early version of this chapter was presented as a paper at The Kennedy School of Government, Harvard University, at a seminar series on 'reinventing science policy', November 1995. An interim version was presented at a session on 'Paradoxes of Knowledge Societies: New Policy Challenges', Joint Meeting of the Society for Social Studies of Science and the European Association for the Study of Science and Technology, Bielefeld, Germany, October 1996. I thank the Pages of Beech Tree Publishing for encouragement in writing it up, and John de la Mothe for patience above and beyond the call of duty in including it in this collection.

PART IV

Emerging Issues for Science, Technology and Governance

CHAPTER EIGHT

International Governance in a Technological Age[1]

Eugene B. Skolnikoff

The startling pace of change in world affairs that accelerated in the late 1980s signaled the end of many of the central elements of postwar international relationships. Many forces were at work leading to the momentous events in the countries of Eastern Europe and the Soviet Union, not the least of them the effects of technological change that profoundly altered the internal situation of those countries and their relations with others. Those developments were emblematic of a much larger story in which advances in science and technology (S&T) have contributed to a profound evolution in national and international affairs.

Elements of that larger story can be seen most vividly in dramatic technology-related developments that have had global consequences: the deployment of massive strategic nuclear forces, the nuclear accident at Chernobyl, computerized financial markets that allow currency transactions estimated in the trillions of dollars per day, and the total eradication of the scourge of smallpox. Other elements may be less immediately spectacular, but as far-reaching: the ability to fax documents and to reach any telephone instantaneously and inexpensively on a world-wide basis, the increasing dependence of modern weapons systems on technology and on technological advance, the relevance of technological competence to a nation's competitive position, and the immediacy and global reach of television. Still others serve to underline the more intensive relation between local actions and global consequences, such as destruction of stratospheric ozone due to the widespread use of chlorofluorocarbons, the far-reaching consequences of a disruption in energy supplies or failure of information systems, or the climatic effects of the accumulation in the atmosphere of the effluents of energy technologies.

The international significance of S&T is not only a result of recent advances, though the breadth of interaction and the rapidity of technological change are relatively modern phenomena. Historical examples are countless, ranging from new weapons technologies that altered the fate of nations to industrial technologies that were the basis of revolutions in economies and wealth. It was not only the physical developments of technology that had an impact; the intellectual currents of the Enlightenment, so much a product of the ideas of experiment and rationality that energized the scientific revolution, served to stimulate massive forces for change in the West. The French and American revolutions were products of those forces, as was the design of government in the American Constitution.

Today, given the scale and organization of the scientific and technological enterprises, science and technology have become arguably the most powerful and persistent factors leading to societal change and, necessarily, to change in international affairs. The accelerated commitment of resources to research and development (R&D) during and after World War II has transformed the relatively haphazard climate of invention and scientific research of earlier centuries. There is now in place a formidable, and growing, capacity – a system – for targeting human ingenuity toward the rapid expansion of knowledge, and for the production of new technologies designed to serve perceived or speculative needs. Not only do the products of this system have significant international effects, but its very operation favors the creation of global markets for its products; and, for a surprising portion of the system, international goals provide the underlying motivation for the commitment of R&D resources by governments and even by industry.

The effects of application of technology are so widespread, in fact, and often of such obvious importance in the conduct and evolution of relations among nations, that it is routine for commentators to lapse into florid rhetoric in describing the resulting dramatic change in the international political system. Secretary of State Shultz in December 1987 said: 'Developments in science and social organization are altering the world profoundly – too profoundly for conventional habits of thinking to grasp. History suggests that mankind rarely understands revolutionary change at the time it is coming about', and W. Michael Blumenthal, former Secretary of the Treasury under President Carter, said in his 1987 Elihu Root lectures: 'I believe there is one circumstance which overshadows all else and has set the current period apart: unprecedented, deep and continual technological change ... extraordinarily rapid technological change has thrust upon us new and as yet unresolved problems of governance in the national and international spheres.'[2]

Just what are these new problems of governance? Are they new, or just modified or intensified problems that have always created difficulties for policy-making? In particular, how have the development and application of new technologies affected the nature of the issues decision-makers in government must face and the conditions under which their decisions must be reached? Some overarching effects, drawn from the broad sweep of technology-related international issues can be spelled out. Many are not of unique relevance to international affairs, but apply equally to domestic policy processes. All reflect the truism, honored in rhetoric more than practice, that domestic and international affairs cannot be separated – they are in fact but two parts of one whole.

Time

One of the most interesting implications for governance that accompanies technological change is the greatly enhanced significance of time in the policy process. That is evident at both ends of the spectrum, as the time available for consideration of complex decisions, sometimes life and death decisions, can be vanishingly small, while concurrently governments have to make policy decisions for material issues with time horizons extending to many decades and, in some cases, thousands of years. Both phenomena are essentially new as significant factors in the councils of government; both have proven to be exceedingly difficult to accommodate in policy processes.

The compression of time scales is evident in many different areas, the most dramatic being the implications for command and control of modern strategic weapons systems. The brief time of delivery of highly destructive strategic missiles, 30 minutes or less over intercontinental distances, combined with the massive quantities of data from real-time reconnaissance and warning systems that must be processed to determine whether an attack is in fact underway, leaves little time for considered review of response options. As a result, responses must be largely prepared in advance, all-but foreclosing the exercise of human judgment in the context of the actual situation.

The proposed Strategic Defense Initiative (SDI), were it ever operational (which is not a realistic possibility), would pose the problem even more starkly, for there could not even be the fig-leaf of required referral to the President for decision.[3] Any attempt to destroy missiles during their launch phase would require the firing of pace-based anti-missile weapons within three minutes – the time the offensive missiles would still be in the atmosphere (though even that could be further reduced). That would necessarily mean a system with automated, procedural responses to electronic signals, allowing no concurrent human evaluation or intervention.

The problem is evident in conventional weapons technologies as well, as the increased speed of many weapons systems, the much more complex information environment, and the firepower of individual weapons can severely limit the time available for considered decisions.[4]

Time compression is also evident in economic sectors where, for example, the growth of global 24-hour computerized financial markets has resulted in a trading system involving literally trillions of dollars that is dependent for its operation on computerized evaluation of market developments and pre-programmed trading decisions. In this situation government must be able to respond with great (and uncommon) alacrity if incidents, such as extreme price volatility that led to panic sell-offs, arose that would have serious economic consequences. Such incidents could themselves be a product of the instabilities introduced by computerized program trading, or by machine (computer or communications) failure.

The rapid global spread of information through communications networks and the media leads to a different example of time compression, creating pressure for quick reaction to important developments wherever they may occur. Officials might prefer time for contemplation or analysis, but an immediate response is often a requirement on the global stage that emerges instantaneously for high-value issues. The absence of response or a deferral can have high costs, including loss of control of an issue. In any case, the time available for consideration of important policy issues is often severely limited as interested parties at home and abroad quickly become informed, and add their voices and actions to the requirement for hasty decisions.

The need for rapid response further increases the reliance on the first information received about an event, information that is likely to be incomplete and probably woefully inaccurate. This enhances the risk of emotional or irrational responses on the part either of decision-makers or the public, or both.

The situation is exacerbated by the broadening of the agenda in which senior officials are involved. The time they have to devote to any single issue becomes steadily more constrained, and has become a steadily scarcer resource in the policy process. This form of scarcity is typically little understood outside government as senior officials are expected to be knowledgeable about, and have time for, any and all issues that arise.

The late Harold Macmillan, a former British Prime Minister, put it with his usual style when he decried what he called this 'instant politics', the assumption, because of radio and television, that leading officials should be expected to comment immediately on any event.

> Every day they talk on television – every day, immediately an event comes. There is no time for thought, no concentration; you have to come on at once or else you're thought to be out of it. ...You must get away, you must read a book. ... You must be quiet.[5]

One consequence of this pressure of time in the policy process, especially on those issues that also pose the greatest dangers and the greatest likelihood of leading to crisis, is the tendency to centralize foreign and security policy responsibility within governments. There may no longer be the time available for extensive consultation, nor the willingness of a head of government to entrust responsibility to a subordinate official, when considerations of time may make that equivalent to delegation of authority. Moreover, it usually will be, or will seem to be, more efficient, faster and safer to conduct policy from a central office when such rapid response may be required. Hence the tendency for security policy to be centered and conducted from the White House or the counterpart offices of other world leaders (Sorenson, 1987/8: 236).

Another consequence is to increase the reliance on machines for the collection of data, for its analysis, and for the transmission of a response. Technology is the primary means for dealing with the problems of compression of time (as well at their cause). That is, technology makes it possible to reduce, process, and analyze information and to implement decisions much more rapidly than in the past. The dependence on machine-generated data and analysis thus grows stronger, with added vulnerability to the unavoidable imperfections of the equipment, and to the built-in biases and assumptions of those who design or program them.

At the other end of the spectrum of time, governments are grappling with major issues which require consideration of consequences and of the viability of commitments extending far into the future but for which decisions must be made today. The danger of global warming is a vivid example, as measures to prevent or delay its arrival would have immediate effects on national economies, while the consequences of warming that does take place would not occur in the next year or two, nor in the next political cycle, but over generations. The disposal of radioactive and highly toxic waste is an issue of a different kind that also requires policies today that must take account of the commitment for safeguards over thousands of years.

The ability to weigh appropriate policies in these situations is hampered by many difficulties, not the least of them the procedures for comparing investments intended to produce benefits over varying times in the future. Cost-benefit analysis using any reasonable discount rates to account for the time value of money makes the present value of expenditures whose benefits will appear 30, 40 or more years in the future usually uncompetitive with expenditures that have shorter-term payoffs. (Rothenberg, 1992; Schelling, 1995; The Synthesis Panel, 1991: 11–12.) The problem is exacerbated by the inherent uncertainty in the estimation of future benefits, while the immediate expenditures to reap those uncertain benefits must be compared against the opportunity costs – the alternatives whose benefits are nearer and clearer.

Alternative approaches have been considered, for example the use of a 'social' discount rate in cost-benefit analyses that could be zero or even negative. Given the intense competition for resources to meet immediate social needs however, governments would not likely be willing or politically able to introduce procedures that would as a general matter bias decisions toward future and against current needs.

Thus, the significance of time in the policy process in governments has taken on new and particularly difficult dimensions. This must, however, be kept in perspective. Time has always been a scarce resource in the making of policy, always a constraint on deliberative formulation of policy. Nor is it all that unusual for governments to make decisions and substantial investments (dams, for example) for which the benefits would be realized well beyond current generations or beyond the terms of current office-holders.

The difference now is that at the compression end of the spectrum, it is not only that time is relatively an even scarcer resource than it has been, but that more, and more consequential, decisions are physically required to be made in a very much shorter period of time than in the past. At the other end, long-term effects must be explicitly taken into account in a larger number of cases, many with consequences that would be felt on a global scale.

The disparate time scales of technological and societal change also add to the pressures on decision processes. Technology evolves faster than the social context and regulatory environment in which it is embedded. Even with efforts at technology assessment that attempt to anticipate the broad, unplanned effects of technology, the social setting will always be largely reactive to technological change as it appears. The result is that policy processes are continually being faced with new unanticipated situations stimulated by technological change, and new requirements for accommodating its effects. The capability for rapid response to change becomes a significant element in a policy process, a process that is, however, inherently structured for a much different and slower rate of change.

Technical content of issues

The inextricable and pervasive involvement of scientific and technological factors in so many issues in international affairs means simply that those factors must be included sensibly in policy processes. How relevant they are to a specific issue, whether their details are central to its consideration or simply part of the general framework, will vary from issue to issue and within issues. In this respect, they are no different from other aspects of policy: an element that must be considered, but the importance of which depends on the issue and on the particular questions being asked.

The obvious lesson to be drawn is that policy processes must include effective representation of the technical aspects of issues. Most governments have found that to be much harder to accomplish in their foreign policy establishments and at senior levels of government than might be assumed.[6]

One of the many difficulties encountered in all policy areas, but particularly it seems in the foreign policy domain, is the limited understanding of science and technology, and particularly technology, on the part of many of those in foreign policy positions. This, too, is not a new problem. Whether it is due to the recruitment and promotion patterns for the foreign service, the traditions and nature of their roles, the necessarily limited technological resources at their command, or other factors, the result is often limited ability to consider anything other than the surface technical characteristics of issues as presented by the technical mission agencies of government. In turn, those technical mission agencies typically have only limited understanding of the broader foreign policy aspects of an issue, and moreover have a stake in presenting information so as to support their own policy objectives.[7]

The result frequently is implicit deferral by foreign offices to the policy perspectives of technical departments, misunderstanding by foreign policy officers of the dynamic nature of technology and of its potential responsiveness to policy intervention, and, ironically, often a highly exaggerated view of what technology might be able to contribute toward achieving policy goals.

The overvaluation of technology's potential contributions is a relatively new phenomenon, and perhaps has become at least as much of a problem as not understanding its rule in policy. The cause has obvious roots. Advancing technology is not only impressive in contributing to the expansion of mankind's wealth and capabilities, but increasingly science and technology have been successfully used to multiply options for dealing with recalcitrant issues. Whether it be substitutes for scarce resources, or more productive food strains to stave off famine, or means for monitoring military movements to avoid surprise, or lunar landings to refurbish international prestige, required technologies have been developed apparently on call.

But, all desirable technologies are not feasible, nor viable in society even if technically feasible. More important, whether technology is a useful contributor to coping with social issues of significance or not, it is certainly never the sole determinant of the outcome. Science and technology offer greater flexibility and a broader range of options for policy than in the past, but excessive optimism about the capability of technology to solve social issues must be severely tempered. Technology can contribute, but social problems require social solutions.[8]

It is not necessary that non-specialists in foreign policy positions become scientists or engineers in order to be able to work with the scientific and technological aspects of issues. Rather, a well-crafted program for non-specialists that offers the same level of exposure that is encountered in good programs of economics or industrial management or public health can provide the basic tools that are needed, including an appreciation of how to approach the scientific and technological factors, how to understand them in interaction with other elements, and how and where to obtain further information when it is required.[9]

One increasingly common development does, however, pose large problems for laymen and even for those with technical training in a policy process. It is the need to represent the technical aspects of an issue in a complex, usually computer-based analysis that is likely to be opaque to all but the programmers. The internal structure and assumptions that must go into these analyses are critical to the conclusions reached, but are not inherently independent of the personal views of the analyst: they require the exercise of personal judgment in the selection of the important variables, in the simplification always required, and in the analytical methodologies employed. Policy conclusions or options are unavoidably vulnerable to the shading that results from the intended or unintended biases of those doing the analyses, but the biases are difficult to discern because of the technical complexity of the analysis. Whatever the significance of the bias to the conclusions, policy-makers are dependent on inputs that they cannot personally understand or evaluate, made more problematic by a tendency for undue credibility to be accorded to computer outpost and to quantitative, rather than qualitative, analyses (Skolnikoff, 1985: 124–35).

Another somewhat paradoxical aspect of the technological nature of many issues is the need to understand and deal with uncertain data and conclusions. Uncertainty figures often and importantly in almost all the issues involving science and technology and is, of course, a necessary part of all policy processes. The quantitative nature of scientific and technological aspects of policy, however, tends to convey a sense of precision that is often wrong or misleading. Moreover, as more global-scale issues emerge, the levels of uncertainty that must be confronted in policy processes are likely to become larger and more pervasive than they have been, greatly complicating the ability to develop collective international action.

Coping with uncertainty and with 'inaccessible' computer-based analyses are not wholly new problems of governance, being particularly evident in security and energy matters for some time, and increasingly so in regulatory and environmental subjects. But, the significance of the technological aspects of a growing number of politically salient issues makes this a more common problem, especially at the higher echelons of government as more domestic and international issues with major scientific and technological components reach those levels.

There is no 'solution' for this problem of dependence on technical analysis. And it is made worse by the concomitant flood of information pouring into governments and decision-makers. This increasing generation and flow of information on a worldwide basis itself creates major problems for policy-makers. Even the application of technology to make more manageable the information must be considered in the formulation of policy does not overcome the problem of information glut, for technology cannot, in advance, separate the important from the merely relevant without sacrificing to the programmer the decision as to what *is* and *will be* important.[10] Once again, the reliance on machines for selection and choice, or to the values of the programmer of those machines, is an increasing, and increasingly disturbing, phenomenon.

One of the more startling examples of dependence on technical analyses carries the role of science and scientific research to unusual lengths. The forecast that global warming will result from increasing concentration of greenhouse gases in the atmosphere is based on elaborate models of global atmospheric circulation that are analyzed by means of repeated computer runs. The models are complex, highly technical, limited in number, and include assumed relationships among important elements that are not yet well understood (e.g. ocean absorption, role of clouds). Yet, the predictions are the underlying justification for the issue being on the global political agenda, where it could result in agreements that would have major impact on national and global economic and political interests. The example is symptomatic of the increasing dependence of many international issues on science.

There is also seen in this subject of global warming a quite unusual role for the scientific community. The computer modeling and analyses that are the basis for the concerns over warming are the product of the work of the community of atmospheric scientists, working to the best of their abilities and judgments with the latest but still insufficient tools. They must simplify the enormous complexities of the atmosphere as they construct their models, which even then are not accessible to the layman, and present results that cannot be more than uncertain predictions, the importance of which are in turn matters of judgment. (Skolnikoff, 1990: 77–93.)

The responsibility shouldered by those scientists in their work and, at least as important, in how they present it, is enormous. By extension, the organizations of the international scientific community that are now heavily engaged with the subject and its implications, also bear major responsibility. There are massive policy implications dependent on what they produce and when. Perhaps the reliance on the scientists of the Manhattan Project represented a similar responsibility for that time; it is not evident any other example comes close.

The dependence on science and scientists in the case of global warming may be unique, but the increasing relevance of science and technology to international policy issues mandates an important role in policy processes for scientists and engineers. They may not be the 'new skilled policy elite' that will dominate policy, as was once predicted, but they have a responsibility to be as much a part of the process of integrating technology with other relevant factors as do the individuals without technical backgrounds who must learn how to cope with the technological factors (Wood, 1964: 41–72).

Role of foreign ministries

All of these issues clearly have an impact on the role of foreign ministries in governments usually, as noted, to enhance the importance of the more technical agencies in the determination of policy. This consequence is exacerbated by the intensified integration of economies and societies, one of the more dramatic international effects of technological change. This growth of interdependence necessarily means large increases in the sheer volume and breadth of relationships between and among states. This has the ancillary effect of further undermining the ability of foreign offices to maintain their traditional dominance in the making of foreign policy. The portions of government formerly concerned almost exclusively with domestic affairs – agriculture, industry, health, environment, education, for example – are increasingly engaged in matters that directly affect international relations and a nation's foreign policy. Those 'domestic' departments have become legitimate players in the making of foreign policy; foreign offices no longer can automatically dominate the process as it was once assumed they should do.

The result is that foreign offices are now typically but one among many ministries on a wide variety of foreign policy issues, the actual situation determined by the particular issue and personalities involved (Skolnikoff, 1967: 249–98). The spread of involvement in international affairs across governments cannot be halted or reversed; it is a characteristic of the intensely interdependent world that is continuing to develop. Attempts to reassert the authority of a foreign ministry, typically by moving decision-making processes higher on the organization chart in order to give a Foreign Secretary greater authority, cannot be continued indefinitely without overburdening the available time of senior officials. Ultimately such moves simply recreate lower-level problems at a higher place in the government hierarchy.

This general problem is exacerbated in the US where power is divided between the legislature and executive. The same difficulty of formulation and coordination of policy across widely varying subjects of international affairs is recreated in the even more chaotic policy processes of the legislature, thus making it even harder for the US to develop coherent policy positions.

The dispersion of involvement among agencies in international issues would be enough on its own to bring about the gradual loss of dominance by foreign offices. The role of domestic technical agencies is enhanced, however, by the increasing relevance of the technical aspects of issues that play to their greater knowledge of relevant technologies. Technology is never the sole determinant of an issue, but in the policy process knowledge is power, especially if there is significant imbalance among participants in their understanding of a key element of an issue. When a technological aspect is critical, the greater technological knowledge on the part of technical agencies makes it possible for them to bias their representation of those aspects in ways that support their view of the international interests of the nation over that of the foreign ministry. It is difficult for a foreign ministry to counter, or sometimes even to know there may be alternative interpretations.

International cooperation

The requirement for collective action to achieve common purposes or to attack common problems is also a natural consequence of increasing integration of economies and societies that owes so much to technological change. The need for collective action is bound to grow as integration increases and becomes an ever-more prominent aspect of international transactions among nations and as more global-scale issues emerge. Much cooperation takes place in the private sector without extensive government involvement; a substantial and growing portion requires commitment and agreement of governments, more frequently than in the past on subjects of substantial political and economic significance.

The policy process that determines the policy of nations toward collective action, however, continues to be entirely national in structure, giving representation to domestic interests affected by the issue, and only indirectly to foreign or international interests. At the least, the process skews the consideration of international issues by underrepresenting the stakes of other affected parties or of the broader global community; at worst, it results in a parochial, nationalistic, view that gives little weight to broader concerns.

In these nationally based processes, a politician or official finds it easier and safer, in fact rational, to choose to take short-term gains through unilateral domestic actions, rather than to gamble his or her own interests on the usually messier problems of international coordination or cooperation. Moreover, for nations with open political structures that provide extensive opportunities for intervention by domestic interests in the policy process, as in the US, the result is to make the development of consistent international policies, the commitment of resources for international purposes, or the acceptance of onerous regulations required by international agreements, much more difficult when there is perceived conflict with domestic needs and priorities.

This strong national perspective that quite naturally persists in the nation-state system, no matter how much erosion there has been around the edges, is what makes problematic the prospect of truly extensive uncoerced collective action if that would seriously constrain national autonomy. Collective action may be necessary if a global issue, for example global warming, turns out to be a catastrophic threat to the planet; but even in that case, the alternative of assertion of hegemonic coercive power by one or a few states may be the more effective, and thus more likely, route to dealing with the issue.

A particularly important, and steadily more pressing aspect of international cooperation, involving both the public and private sectors, is the increasing dependence of society on large technological systems. This is a development that stems primarily from the economic and performance imperatives of new technologies that benefit from (or require) large, international systems to be operated efficiently. Such systems affect problems of governance in many obvious ways due to their interaction with other issues in national affairs (e.g. regulatory, economic, fiscal policies). The most unusual problem, however, and perhaps the most challenging, is the need to have the capability to respond to the inevitable crises that will arise from the vulnerability of the systems to disruption.

This is a generic requirement for governments that is relevant to any large, complex system, but particularly to those that are heavily technological in nature (e.g. computer information systems, electric distribution grids, telephone systems). Measures to protect against system breakdown are necessary, but perfect protection is not possible (short of abandoning the technology altogether, as many advocate for nuclear power). Breakdowns are sooner or later inevitable. The policy processes of states, singly or in consort, must be prepared for breakdown and capable of dealing with their effects, including longer-term consequences, without knowing in advance where they will occur, or what their dimensions will be. That is not a simple challenge for a state coping with an internal threat; it is that much more demanding in an international environment among sovereign states. It is possible that eventually, if the crises resulting from systems breakdown are sufficiently costly, new international mechanisms with substantial emergency (at least) authority over activities within nations may prove to be necessary and acceptable, even to 'sovereign' states.

One particular form of international collective action in science and technology is of obvious special interest. International cooperative programs, requiring varying degrees of joint planning, commitment of human and financial resources, and sharing of results, have been growing more common since the wartime years. The reasons are not hard to discern: important issues that cannot be tackled in one country alone, increased costs of research that put a premium on sharing, greater competence in science and technology throughout the world, stronger international mechanisms for coordination of planning and operations, even, political benefits that can be harvested from cooperative projects. (Keynan, 1991: Wallerstein, 1984.) The number of cooperative projects is large, and is likely to increase as the importance of some subjects, such as the global environment, that *require* international research cooperation become more politically salient.

For all these reasons, this should be a particularly propitious time for more extensive international scientific and technological cooperation; in fact, the record has been spotty, especially on more ambitious, large-scale, projects. There are many reasons, some of a purely practical nature stemming from the difficulty of meshing differing national systems of budgetary process and project selection. (Skolnikoff, 1996.) The fact that the support of R&D in all countries is determined in a national policy process is the source of many of those practical problems.

But, perhaps the most important reason is that equitable international collective action requires an effective and committed leader. The US has played that role responsibly since World War II as it maintained effective technological dominance. Now, that dominance is gradually fading, as the natural result of growth of capabilities of European nations and Japan, and the supportive policies of the US itself. (Kennedy, 1987; Nye, 1990, see discussion in ch. 4.) The US, though the strongest nation across the board in science and technology, is nevertheless only one among equals in any given field, sometimes ahead of others, sometimes behind.

The US has found it difficult to reflect this diminished technological dominance adequately in its policies toward cooperation. Several large cooperative programs have been undertaken with the implicit assumption (in the US) that the nation would on its own determine program scale and objectives, and others would join in later almost as subcontractors. Some collaborations were treated cavalierly, with an agreed project restructured without consultation, or even canceled without notice. The country has earned an uncomfortable reputation as an unreliable partner in scientific cooperation, though as resources become tighter, this attitude is moderating. (Keynan, 1991: 41–60.)

In fact, for collective action to work successfully among independent nations, a much more forthcoming attitude is required. There is usually a need for a leader to initiate and spur cooperation, but there then must be genuine interaction, starting with the determination of objectives and initial planning and proceeding through the many steps of implementation. And, there must be the certainty that commitments will be kept. Those lessons may be difficult to accept for a nation that has been used to dominating programs with money and knowledge, which has a budget voted on an annual basis, and a policy process that emphasizes domestic interests. But, international cooperation involving the US will always be flawed, to the great cost of the US and of the international community as a whole given the great size and competence of US science and technology, if those lessons are not somehow reflected in improvements in the policy process.

International organizations

One of the consequences of increasing integration of economies and societies aided and abetted by technological change is the need for international institutions that are able to cope with the international consequences of technological change or are necessary to realize the benefits of technology. The expansion in the number and role of international governmental organizations (IGOs) of both limited and universal membership in this century and particularly in the last several decades, reflects this need. (Jacobson, 1984: 30–58; Zacher, 1991.)

Permanent IGOs were first established in the nineteenth century, and have grown rapidly in number and scope in parallel with the growth of interdependence. Harold Jacobson (1984: 47–52) listed 621 IGOs in 1980 (there were at least ten times that number of international NGOs – international non-governmental organizations – not including multinational cooperations) (Archer, 1983: 130; Zacher, 1991: 6).[11]

The large majority are organizations created to carry out specific purposes, often strongly technology-related, such as regulation of the radio-frequency spectrum (International Telecommunications Union), control and eradication of disease (World Health Organization), control of atomic energy (International Atomic Energy Agency) or regulation of commercial aircraft (International Civil Aviation Organization). All, whether narrow in purpose or designed to serve larger political, security or economic goals, bear some direct or indirect relation to science and technology, just as does international politics as a whole.[12]

For a time, the organizations were seen by many as precursors of an important change in the international political system. It was thought that delegation of authority and power from nation-states to international organizations in order to carry out 'essential' international functions would over time lead to erosion of the independent power of nation-states and to gradual political integration in a larger political entity.[13]

A considerable measure of such delegation has taken place, but not in powers that seriously derogate from the independent authority of states. With the important exception of Western Europe, little political integration has taken place as nations have remained jealous of their prerogatives – even more so as they feel some have been diminished – and have tended on the whole to accept internationally based regulation and management only when it was clearly beneficial and vital national interests were not at stake. National policies toward IGOs are conditioned by the same factors as policies for international Cupertino in general, making them hostage to domestic interests in a nationally based system. When IGOs serve or benefit domestic interests, official attitudes are supportive. When they interfere with, or complicate, achievement of domestic goals, official attitudes tend to be hostile, or at least indifferent.

It is nevertheless true that the role and functions of IGOs are a considerable factor in the gradually increasing constraints on national autonomy that all states have experienced as interdependence has intensified. And, it is also true that any organization, no matter how circumscribed in its charter or oversight, develops some measure of autonomy and independence over time. That is equally true of IGOs; they are in many ways not the totally servile eunuchs of nation-states they are often assumed to be, but have, to varying degrees, developed considerable autonomy and independence (Skolnikoff, 1972: 132–48). Also, they go through a process of learning and adaptation as their framework evolves and their primary purposes gradually change.

IGOs not only perform essential functions, but those with a more technical purpose are in fact quite effective in their operation, more effective than is often assumed (Skolnikoff, 1972: 158–66). They have to be. The requirements for managing an increasingly integrated world could not have been met without them, and their existence and performance imply that nations recognize they can not achieve their national goals, domestic or international, without an international structure that includes a growing number of reasonably autonomous and competent multinational organizations.

An obvious question is whether the growing need for effective collective action that accompanies technological change will result in steadily increased delegation of authority, perhaps leading to a more substantial evolution in the role of IGOs sometime in the future. In 1946, the serious attempt to internationalize control of the atom foundered on the rock of big power politics, even in that time of strong and, as it turned out, idealistic hope for the future authoritative role of the UN (McGeorge, 1989: ch. 4). Perhaps tomorrow, threats of catastrophe, whether ecological or economic, will make it imperative and acceptable to delegate substantial economic and political authority to international organizations.

It is also possible that, rather than accepting the growth of authority of IGOs at the expense of nation-states in the face of crisis, one or a small number of states will take matters in their own hands and achieve the necessary actions through more traditional means of exerting power.

A possible harbinger of changing attitudes toward international institutions, however, may be the demonstration, when Iraq invaded Kuwait, of the use of the UN to mount a genuine collective response to aggression. That may have been possible only because of the clear-cut violation of international norms, and the particularly egregious and barbaric behavior of Iraq's President, Saddam Hussein. The more important precedent for the long run may well be the subsequent postwar acceptance of an intrusive monitoring and inspection role for the UN in Iraq as they search for weapons of mass destruction. That represents an unprecedented mandate for interference in normal sovereign prerogatives, agreed to in extraordinary circumstances, but of potentially great importance for the future role of the UN.

The development of the European Union represents a rather different example. That assemblage of reasonably like-minded states of roughly equal wealth has been making tortuous, but steady, progress toward aggregation of responsibility in the European Commission, with substantial ceding of authority to that body. The extent of the progress that has been made is impressive, and does demonstrate that independent nations can recognize the limitations of their nationalism and the advantages of pooling their sovereign powers. The reunification of Germany, the collapse of the military threat from the East, the plans for introduction of a common currency, and the eagerness of additional nations to join, has introduced new conditions for the Union, with presently unclear consequences for the previous momentum toward real unification. The political dimensions, in particular, are uncertain (the Union did not perform well as a unit in the political run-up to the Gulf War).

The experience of the European Union does show, however, that under certain conditions, mature nations are willing to sacrifice major elements of their autonomy when the benefits are seen as substantial and the alternatives sufficiently undesirable. Those certain conditions may, however, be unique to Western Europe.

It is well to note that the expansion of the number of IGOs is far outstripped by the increase of non-governmental organizations (NGOs) operating in an international environment.[14] It is not surprising, for the growth of cross-border relationships that is a part of interdependence, naturally includes increased international Cupertino and collusion among domestic-based interest groups. They have become an important part of the international scene, in no small measure because of their effective use of the fax machine, the media, and other information technologies. In some cases, they have come to play quite prominent and influential roles in the evolution of policy. Many are organized around environmental issues, and are now a visible presence at most international environmental negotiations.

The extent of the policy influence of NGOs is not uniform among organizations or subjects, nor can it be estimated responsibly without extensive study. However, over time their influence is likely to grow as they accumulate more members, and exert pressure on more governments. They are likely to be a force that lends nations to turn more readily to international bodies on transnational and global issues, and to accept collective action.

Their role, however, is not an unalloyed positive. NGOs, being private, are not representative organizations; the participants, often the self-appointed leadership, set their own agenda. They often have limited resources and have difficulty following through to the implementation phase of agreements. Moreover, they are typically organized around narrow issues, sometimes only a single issue, that gives little room for interaction and compromise with the agendas of other NGOs or of governments. And, their usually narrow and committed advocacy role has sometimes resulted in questionable use of science and scientific evidence.

Perhaps the largest international impact of NGOs will prove to be in the propagation of a common act of values toward contentious issues in the environmental, social or population domains. This makes some developing countries quite worried. Representatives of developing countries at international conferences will often voice off-the-record concern that NGOs are dominated by industrial country values, so that what is being advocated, and increasingly accepted internationally, will in fact be inimical to developing country interests.

Public involvement: political fragmentation

One other implication for governance worthy of much longer analysis, but not peculiarly international, is the consequence of increased public involvement in policy processes that accompanies the greater availability of information to the public at large. Clearly, the extent and significance of this will vary from country to country, and from one political system to another. The efforts in Eastern Europe and the Soviet Union proved to be of great political importance in leading to the change in the governmental structures of those countries. In the West, the easier availability of information has been a factor tending to support increased public intervention in the policy process through organized interest group activity, often focussed on single issues, and the concomitant fragmentation of electoral politics.

It is not only increased public involvement *in* policy processes that affects those processes, but also the increasing ability of groups forming within or among nations to bypass established lines of authority and operate on their own. The multiplication of subnational and supranational actors serves to constrain the autonomy, and often the effectiveness, of governments; in turn, that serves to encourage and empower even larger roles for those non-state actors.[15]

Conclusion

The many effects on the nature and problems of governance in the international system that are a product of technological change occur only gradually, in principle leaving time for policy processes to adjust. However, as with all large systems, adjustment comes slowly, with large time lags, and often only in the context of a crisis of some sort that grows out of the mismatch between the need for change of policy and the actual decision process. Technological change makes timely improvement of governmental processes even more difficult than it might otherwise be, in part as a result of the rapidity of change and in part because it is not possible to protect fully against some of the new problems introduced in the policy process such as dependence on machine processing of information for urgent decisions, unequal access to the technological elements of issues, or unpredictable breakdown of essential technological systems.

It is not novel to recognize that the effectiveness of governance, domestic or international, will always be lacking to some degree compared with what might or should be possible or with what is needed. The problems introduced by technological change, however, increase the gap and make it harder to achieve the level of effectiveness required by a more complex, technological world. It is thus one more aspect of the many consequences of rapid technological change that all are experiencing.

Notes

1. The ideas for this chapter and some of the text are drawn from the author's *The Elusive Transformation: Science, Technology, and the Evolution of International Politics* (Princeton University Press, 1993).
2. Address by Secretary of State Shultz before the World Affairs Council, 4 December 1987, published in Department of State, Bureau of Public Affairs, Current Policy Number 1029; Blumenthal, W. M. (1987/8) The World Economy and Technological Change. *Foreign Affairs*, **66**, 3, 531.
3. See Skolnikoff, *op. cit.*, pp.65–71, for a detailed discussion of the unreality of the Strategic Defense Initiative.
4. In post-mortems after the Gulf War, Pentagon officials expressed 'amazement ... about how quickly warfare is being overtaken by technologies that compress time and therefore affect human judgements' (Patrick E. Tyler, 'Pentagon reassesses the vulnerability of its tanks', *New York Times*, August, 1991, p.A10).
5. 'An Evening with Harold Macmillan at the Wilson Center', Woodrow Wilson Center for International Scholars, Washington DC, 1986, p.8.
6. There is a substantial literature on the interaction of science and technology with foreign policy, and on the roles and problems of scientific experts in policy formation. See particularly Sanford Lakoff (1966) *Knowledge and Power*, New York: The Free Press; Joseph S. Szyliowicz (ed.) (1981) *Technology and International Affairs*, New York: Praeger; Commission on Organization of the Government for the Conduct of Foreign Policy (1975) *The Management of Global Issues, Appendix B*, Washington, DC: US Senate, June 1975; Eugene B. Skolnikoff (1967) *Science, Technology and American Foreign Policy*, Cambridge, MA: MIT Press.
7. The general problem of scientific and technological understanding in foreign ministries, and its implications, is discussed in Skolnikoff, *Science, Technology, and American Foreign Policy*, pp.249–98. A report of the Carnegie Commission on Science, Technology, and Government analyzes in detail the situation and needs in the US Government: 'Science and Technology in U.S. International Affairs: Challenges for 21st century Governmental Policy and Organization', New York: Carnegie Commission on Science, Technology, and Government, 3 October, 1991.
8. It is worth noting that it is not only those without technical training who are vulnerable to overvaluing technology's potential contribution. Scientists and engineers have proven to be

culpable as well. The most egregious example is the overselling of SDI by a number of scientists. The attention accorded the well-orchestrated promise of unlimited energy from 'cold fusion' provides another illustration (Close, F. (1991) *Too Hot to Handle*. Princeton: Princeton University Press).

9. Carnegie Commission, 'Science and Technology in U.S. International Affairs' (see note 7).

10. The Revolution in Information and Communications Technology and the Conduct of U.S. Foreign Affairs. Report of a Workshop, 14–15 September 1987. Washington, DC: National Academy Press, 1988, 31–55.

11. Different authors have different absolute numbers, depending on definitions; the proportions are similar, however.

12. There is an extensive literature on international organizations, their origins, functions, and political relationships. See particularly Jacobson, 1984; Mitrany, 1966; Haas, 1964; Archer, 1983. For an analysis focussed on the role of science and technology in those organizations, see Skolnikoff, 1972.

13. Substantial theoretical schools were based on this general idea, growing largely from the work of David Mitrany (*A Working Peace System*) and, later, Ernst Haas (*Beyond the Nation-State*) and Joseph S. Nye ((1971) *Peace in Parts: Integration and Conflict in Regional Organization*, Boston: Little, Brown).

14. Zacher estimates that international NGOs in fact grew much faster than IGOs in the postwar period (Zacher, 1991: 6). Archer (1983: 130) estimates close to 10,000 international NGOs by the end of the twentieth century.

15. James N. Rosenau (1988) discusses this point in 'The state in an era of cascading politics', *Comparative Political Studies*, 21 (1), 13–44.

CHAPTER NINE

The Governance of Military R&D

Judith Reppy and Philip Gummett

Introduction

The management and control of government programs for military research and development (R&D) have dominated thinking about the governance of R&D in a number of leading industrialized countries for most of the post-World War II period. The reason is quite simple. For many years, national security was the principal rubric justifying government spending on R&D, and military spending dwarfed all other categories in national R&D budgets. This was true above all in those countries like the United States, the Soviet Union, Britain, and France that maintained large military forces, including nuclear forces, across the whole spectrum of military technology. The mission-oriented 'style' for government R&D in these countries can be traced directly to the central place they have given to military R&D. (Ergas, 1987: 191–245.) As a result, most of the standard government practices for planning and overseeing large R&D projects have evolved in the context of military projects and the problems they pose for accountability and efficiency. And, while the bulk of military R&D funds has been devoted to engineering development activities, military spending on basic research, particularly in the United States, has been significant in many fields and so has played, at least implicitly, an important role in government policy for fundamental science, as well.

Government policy for military R&D is everywhere centralized at the national level, even in those countries that have decentralized structures for civil technology, and legislative oversight it typically constrained by secrecy and deference to military expertise. The risks are perceived as high: better to err on the side of doing too much than too little when national security is at stake. Compared to other areas of science and technology there is little public participation or debate on government policy for military R&D, the main (and important) exceptions being the anti-nuclear campaigns in many countries and the anti-ballistic missile (ABM) and Star Wars debates in the United States. Even in these campaigns, however, the focus has been largely on the deployment rather than the development of weapons. Military R&D, although large compared to other areas of science and technology, is small compared to total military spending, and thus relatively protected from controversy in debates over defense programs.

Military R&D has other special characteristics, beyond the sheer volume of resources that has been lavished on it and the legitimacy that it enjoys as an element of national security. In the broadest sense it responds to the requirements generated by military doctrine (which in turn is influenced by the technological capabilities created by R&D), but the relationship between doctrine and technology is mediated by the specific environment in which military R&D is performed. That environment is marked by secrecy, inter-service rivalry, government bureaucracy, alliance politics, and constraints imposed by international arms control agreements. The defense market it dominated by a single buyer that traditionally has valued high performance over cost considerations. It is also notable for the close relationships that have developed over time between the government buyer and its suppliers in the defense industry.

In pursuing military R&D, governments face a classic principal–agent problem, characterized by asymmetric information and high levels of risk (Guston, 1996: 229–40). Solutions to this problem have ranged from reliance on in-house government laboratories, to hybrid public–private organizations such as the federally funded R&D centers in the United States, to contracting with private industry or with universities. Each of these alternatives, however, presents particular problems for devising appropriate incentives and controls. In-house solutions offer the government the greatest control, but as part of the government bureaucracy, government laboratories are handicapped in managing R&D activities, which, many would argue, are best pursued in an atmosphere of minimum external control. Private contractors, by contrast, have greater freedom of action – for example, in setting salary scales for scientists and engineers – but the government must protect itself against the problems of 'moral hazard' (i.e., being unsure how closely the agent is sticking to the principal's goals after the authority to proceed has been delegated) and 'adverse selection' (deciding which of the possible agents most closely shares the principal's goals) that are the result of its imperfect knowledge about contractors' capabilities and performance. Indeed, the more that the government relies on non-governmental organizations (and the less, therefore, that it relies on its own research establishments), the weaker becomes the base of expertise upon which it can draw when evaluating potential contractors. In other words, a government needs some in-house technological capacity in order to be a smart customer, a task that has become increasingly difficult as it has become impossible for any one actor to hold all the knowledge needed to make a decision.

In military-technological matters, western countries have generally relied on a mix of institutions, varying over time, with a decided trend toward increasing reliance on private contractors. In these circumstances the choice of contract form becomes a primary means of managing the government–contractor relationship. Fixed-price contracts, in which a fixed price is negotiated at the beginning of the contract, place the risk on the contractor, but also entitle it to resist pressure from the military customer to alter the specification of the task once underway. At the other end of the spectrum, cost-plus contracts – which have generally been seen as more appropriate for R&D projects because of the high degree of technical uncertainty involved – place most of the risk on the government, which agrees to reimburse the contractor for all of its costs, plus a fixed fee. Cost-plus-incentive contracts are a third type, lying between the two extremes. The important point is that whatever type of contract is chosen there are unavoidable trade-offs. The government cannot escape the dilemma inherent in wanting to minimize both cost and risk while providing incentives to perform the desired R&D. In practice, defense contracts are supplemented by a vast array of certification procedures and post-contract audits designed to guard the public interest against contractor malfeasance, providing yet another trade-off, that between increasing accountability and reducing the burden to the contractor of government paperwork.

Contracts are also supplemented by a web of informal links that are best captured by a network metaphor (Gummett and Reppy, 1990: 287–303; Rhodes, 1996: 652–67). Defense contractors and government officials have numerous ways of trading information and ideas outside of their formal contractual relationships, ranging from open lobbying by companies for particular programs to less visible exchanges between technical staffs. They are facilitated by career patterns in which military officers often retire to company jobs, and company officials are selected for civilian jobs within defense ministries. These network relationships can be viewed as one possible solution to the principal–agent problem in that they provide channels for information flows and building trust and so mitigate the problems of moral hazard and adverse selection. In another sense, however, they undermine the notional principal–agent relationship between government and contractors because power in networks is typically broadly diffused. Consequently, it is only at the highest level of abstraction that we can continue to talk about the government as a principal able to relate in its agents through hierarchical transactions.

The institutions and practices outlined above persisted for many years. The end of the Cold War, however, invalidated many of the assumptions that underpinned strategic and budgetary choices favoring military R&D, in the process altering many long-established relationships in the military–industrial networks. In particular, in most countries budgets for military R&D have fallen along with total military spending, affecting both government and industrial laboratories.[1] Nevertheless, it is important to stress from the outset that in many countries high technology weapons remain central to military doctrine,[2] and military R&D is still a large and important component of government spending for R&D, one that would command our attention on size alone (see Table 9.1). Recent changes in long-term prospects for military R&D and in R&D-performing institutions raise other important issues for government policy. Indeed, it is not clear whether traditional means of governance for R&D are adequate to the new challenges.

The changing context

Cold War interests and requirements shaped the institutions for military R&D that we have today. To counter the perceived threat from the Red Army, the United States and its allies adopted a strategy of relying on high technology weaponry, in which they held a comparative advantage; the Soviet response in developing its own advanced weapons only served further to fuel a technological arms race. National security requirements justified a high level of funding and an emphasis on high performance developments that was seldom questioned throughout the Cold War. Military R&D programs were an indispensable element in the maintenance of this situation: by performing R&D, defense companies developed new ideas, responded to emerging military requirements, and positioned themselves to reap the rewards of follow-on production contracts. Government management of this complex process focussed on efficiency questions – particularly the question of contractual incentives

Table 9.1 Expenditure of military R&D, selected countries

Country (year)	Government expenditure on military R&D (1990, US $ millions)	As percentage of government expenditure on R&D	As percentage of national expenditure on R&D
US (1996)	32,000	54.7	20.0
UK (1994)	3,200	40.8	14.0
France (1995)	4,800	33.6	16.0
Germany (1995)	1,500	9.1	3.6
Japan (1995)	1,000	6.2	1.0

Source: Stockholm International Peace Research Institute, *SIPRI Yearbook 1997*, Table 7.2 (Oxford University Press)

and the problems of maintaining a semblance of competition in a market dominated by a single government customer – and on responding to a series of scandals centered on particularly costly programs. On the whole it was a remarkably stable environment, one in which the potential destabilizing effect of continuous technological change was tamed through long-standing networks summed up under the rubric of 'the military-industrial complex'.

Even before the Berlin Wall came down in 1989, however, long-term economic trends were threatening the established institutions for military R&D. Technological advance was becoming more and more expensive to achieve, leading to a spiral in which smaller buys of more capable weapons resulted in ever higher unit costs, leading to still smaller buys. Budget constraints coupled with rising costs translated into approval of fewer new programs. In other words, rising military R&D costs had to be amortized over fewer units, a situation encapsulated in the satirical prediction that by the year 2054 the US defense budget would be sufficient to buy only one tactical aircraft to be shared equally among all the military services (Augustine, 1986: 143). By the 1980s the industry in most countries was characterized by substantial excess capacity, at least in the traditional sectors like shipbuilding and aircraft production, and this excess capacity extended to R&D. Thus, there were incentives for defense firms to engage in collaboration to gain market access and share development costs, as well as mounting pressures toward industry consolidation.

A second trend was eroding the established support for military R&D programs. The success of Japan and other newly industrializing countries in contesting markets in high technology civil goods – especially electronics, which was the fastest-growing market in the world – raised the prospect that, by devoting large resources to military R&D, countries like the United States and Britain had sacrificed their competitive position in commercial markets. That is, the opportunity cost of military R&D had become clear. In addition, it was increasingly obvious that in some militarily important technologies – again, especially information technologies – technological superiority was now to be found in the civilian, not the military sector (Gummett and Reppy, 1988; Gummett and Stein, 1997). Governments faced the challenge of revitalizing their civilian technology bases, while the military procurement establishments had the problem of overcoming technological isolation.

Both were under pressure to adopt new strategies for military R&D and to rethink the institutional setting in which the R&D would be performed.

These developments occurred in the context of an even more significant trend: the increasing globalization of the world economy. Whether measured by international flows of trade and investment, the spectacular growth in information flows, or the concomitant changes in institutions and practices, the world economy has become increasingly interdependent. The defense industry has not been immune to these influences, although the strong preference of most governments to protect nationally based capabilities in areas important to national security has affected the rate and extent to which the defense industry has experienced globalization. In the defense industry it has taken the form of increased international collaboration; the spread of military technology through arms sales, offset arrangements, and civilian technology channels; and, as discussed below, convergence in national strategies for military R&D. The number of inter-firm teaming arrangements and joint ventures across national lines rose throughout the 1980s, both within-Europe and transatlantic.[3] Corporate interests are no longer delimited along strictly national lines, and in European countries, at least, it is increasingly difficult to speak of a national defense industrial base (Walker and Gummett, 1993).

Thus, the environment for military R&D was already changing when the end of the Cold War ushered in a new era marked by falling budgets and ill-defined security threats. Military establishments on both sides of the East–West divide have lost their familiar enemies and instead face the task of evaluating and countering a host of new potential and actual dangers, many of them emanating from sub-state actors. The military doctrine and forces built up over decades of Cold War confrontation are largely irrelevant to these new threats; armored divisions or nuclear weapons cannot counteract illegal immigration or the drug traffic and are equally inappropriate in most cases of ethnic conflict. In principle, at least, new approaches to military security are needed, with new doctrines that will, in turn, spawn new requirements for military R&D. Thus, instead of the design of new nuclear weapons, we have programs of nuclear 'stewardship' and 'dismantlement'. In addition to the developments that have greatly increased accuracy and lethality of conventional weapons, we have work on 'non-lethal' weapons, which aim to disable without killing, and talk of 'information warfare'.[4] There is a renewed interest in biological warfare and ways to detect and defend against biological agents.

The technologies involved in these new types of weapons draw heavily on the civilian technology base. This is most apparent in the cluster of technologies that make up the so-called 'system of systems'. The relevant technologies – the sensors, software, simulation capability, intelligence-gathering systems in space and elsewhere, and near real-time data processing capability – are all based on developments in microchips and computers. Reliance on advanced information technology, however, brings with it new vulnerabilities: the Pentagon must now seriously consider the threat posed by computer hackers, who regularly penetrate into the Defense Department's network. Skeptics within and without the defense establish-ment have questioned the effectiveness of the current generation of high technology weapons, generating the predictable response that still newer developments on the horizon will fix the problems.[5] But adjusting to new ideas and circumstances is never easy, and most military R&D continues along traditional lines, albeit with fewer resources.

In the United States, military R&D spending has fallen by almost 25 percent in real terms since the late 1980s and is now only about 55 percent of total government R&D expenditure, compared to a recent high in 1986 of nearly 70 percent (National Science Foundation, 1996). Industry consolidation, particularly in the aerospace sector of the defense market, has resulted in 'rationalization' of R&D facilities and in-house R&D spending (Velocci, 1995: 36–8). In 1996, for example, when Loral was taken over by Lockheed Martin reductions in R&D spending were forecast, including cuts in independent research and development (IR&D) (Velocci, 1996: 43). Overall, the net impact of industry consolidation of military R&D has been to eliminate capacity and reduce the number of separate enterprises bidding for government work. The surviving companies have mostly adopted a 'pure play' strategy, selling off non-defense subsidiaries in order to concentrate on the defense market, so that the industry is both more concentrated and more dependent on defense contracts, at least at the prime contractor level.[6]

Lockheed Martin's plans to cut spending on IR&D at Loral are of particular interest because of the role that this program, in which the government reimburses defense companies for the non-contract or 'in-house' research and development they perform as part of indirect costs, has traditionally played in reinforcing military-industrial networks. On average, the Department of Defense (DoD) has paid for roughly half the cost of R&D projects in recent years, reflecting its share in the companies' total sales. Although the R&D program has been controversial in the past, both the Defense Department and the industry have consistently defended it as a source of new and innovative ideas, and it has long served as an informal channel for information on company capabilities and agency interests and as a means for companies to position themselves to compete for new projects. (Alexander *et al.*, 1989.) The changes in military spending and industry consolidation have, however, altered the incentives for participating in the IR&D program. With fewer dollars being spent on procurement of new weapons, the prize for a successful R&D program is less valuable, and industry consolidation has reduced the number of potential competitors. So there are reasons beyond the elimination of frankly duplicative projects for the new corporate giants to reduce R&D spending. DoD's costs may not fall as much as the level of activity, however, because the consolidated firms are more heavily dependent on military sales than their constituent parts were, and therefore they will recover a higher fraction of their IR&D costs from the government.

The changes in the IR&D program are just one example of the way in which changes in the industry structure and in spending patterns have affected the ability of the US government to achieve its goals for its military R&D program. The emphasis on competition that was the centerpiece of DoD procurement strategy in the 1980s is no longer feasible for many programs; the costs of funding parallel development programs is too high, and there are not enough firms left in the relevant market segments to provide competition. As noted above, the most advanced technology is increasingly in the civilian sector. The US government's response has been to turn to civilian suppliers and dual-use technology, a solution that brings its own problems, as we discuss below.

In Europe, too, falling defense budgets and industry consolidation are altering the environment for government policy for military R&D, but in the context of a very much smaller scale, with a correspondingly heightened sense of vulnerability to competition from the US defense industry (Van Scherpenberg, 1997: 99–122). Inevitably, much of the R&D effort goes to tracking developments from elsewhere, although Europe does lead in some technological niches. Just to emphasize the disparity, the scale of the Western European arms market as a whole is less than half that of the United States, while the Department of Defense contracts of the new Lockheed–Martin–Loral–Northrop company are roughly equal to the total annual expenditure of Britain on military equipment (Bagli, 1997: A1, D2; Morrocco, 1977: 24–5; Stockholm International Peace Research Institute, 1997: Table 6A). The European defense industry as a whole remains oversized (although it has shrunk faster in some countries than others), and there are more prime contractors than in the United States, despite the disparity in size of defense markets. For example, the United States has two prime contractors for main battle tanks and armored personnel carriers (Europe has ten), four shipyards (Europe has fourteen), three contractors for missiles (Europe has twelve), and five for helicopters and aircraft (Europe has nine). As a result, European firms cannot achieve the same economies of scale available to US firms, and costs are higher.[7]

Against that background, the scientific, technological and industrial base of Western Europe, like that of the United States, has, overall, depended significantly upon substantial investments in the defense sector. As in the United States, the end of the Cold War brought profound changes in the perception of the threat and the willingness to spend money on defense. Thus, UK spending on military R&D fell in real terms by about 30 percent between 1986–7 and 1994–5, and was projected to fall still further. As a percentage of gross domestic product, UK defense R&D expenditure fell from 0.49 percent in 1989 to 0.33 percent in 1994.[8]

Understanding the changing European policy context is complicated by the need to take into account not merely the flux in national positions, but also the moves toward establishing some more distinct form of European defense and security identity. A key step in this regard was the establishment of a common foreign and security policy (CFSP) under the 1992 Maastricht Treaty on European Union, which policy included the 'eventual framing of a common defense policy, which might in time lead to a common defense'.[9] Within this general framework, the collective security organization, the Western European Union (WEU), which previously was entirely separate, was defined as being an integral part of the process of development of the European Union (EU) as well as contributing to the solidarity of the Atlantic Alliance (that is, NATO). Among the proposals which, under these new arrangements, the WEU was tasked to examine further, was that of enhanced cooperation in the field of armaments, with the aim of creating a European armaments agency.[10]

This treaty language was the result of a compromise between those who wanted a strong European defense and security identity, largely separate from NATO, and those who wished to maintain NATO as their primary security framework, albeit with a strengthened European component. The WEU entered the picture because of the happy accident of close (though not perfect) correspondence between its membership and the EU members of NATO, enabling it to be portrayed as both the defense arm of the EU, and the European arm of NATO.

The reluctance to surrender competence to Brussels in this policy domain is one reason why the CFSP provisions were kept at the inter-governmental rather than the supranational level, restricting the scope for initiative by the European Commission, and hence severely limiting the possibilities for linking the restructuring of defense R&D, or firms, to the other technological and industrial policy instruments of the European Union. Similarly, moves to establish the European armaments agency, led by France and Germany, were bogged down in a succession of difficulties, and only began seriously to advance in November 1996, with the formation, outside the frameworks of both the EU and the WEU, of the Joint Armaments Cooperation Organization (JACO, known in France as OCCAR – Organization conjointe de coopération en matière d'armement). Its founding members are France, Germany, Italy, and the UK; others may join in due course. It will be responsible for the management of several existing Franco-German programs, and will take over responsibility for a range of multilateral collaborative programs, beginning with a new family of wheeled infantry combat vehicles (known as VBCI in France, MRAV in UK, and GTK in Germany).[11]

The establishment of the CFSP at the inter-governmental level, and of the JACO/ OCCAR entirely outside the frameworks of both the WEU and the EU treaties, is unlikely to be a stable arrangement in the longer term, but it reflects the political inevitabilities of the current debate in various key European states. Hence, while the countries of Europe see themselves as increasingly engaged in a search for common solutions to security problems, their thinking on the whole remains dominated by nationalistic concerns, while the sheer complexity of current institutional arrangements makes it difficult to see how best to move forward. There is undoubtedly more internationalization of discussions about policy in this area, but the extent of international decision-making remains very limited.

At the industrial level, consolidation has also not proceeded as far as it has in the United States, largely because national governments have continued to protect their national champions in order to preserve jobs and some measure of autonomy. This situation, however, may be changing. The French government is now committed to a merger between Dassault Aerospace and Aérospatiale followed by cross-border European consolidation as part of a strategy aimed at creating European companies capable of competing with US companies; this commitment survived the change in government in the summer of 1997, although the timetable has been stretched. Whereas France generally espouses a policy of European preference in defense procurement (de Briganti, 1996: 4, 66), the United Kingdom advocates a policy of full-scale international competition, keeping lines open to the United States as well as to Europe. In both cases, however, there is a clear recognition that a purely national defense industry is no longer feasible.

Nor does it make sense to think in terms of purely national policies for military R&D. Since World War II, Europe has moved through a phase of reliance on the United States for military technology, as exemplified by the large number of licensing agreements in the early years of NATO, followed by a growth in indigenous capability as national defense industries recovered from wartime devastation, and finally a move toward cooperative European programs. Parallel with civilian research programs funded and administered at the European level – e.g., in space and in microelectronics (ESPRIT) – a cooperative program of defense research

(EUCLID) has been set in place. The numerous joint weapons programs and corporate joint ventures that cross national lines all contain an element of joint R&D. The Franco-German Eurocopter helicopter company, formed between Aérospatiale and Daimler Aerospace; the formation of Matra (France)–Marconi (UK) Espace, representing the satellite interests of the parent companies; the fusion of the missile interests of British Aerospace and Matra, announced in August 1996; and the possibilities for further international rationalization in military electronics that appeared to follow the decision of the French government in October 1996 to approve the fusion of state-owned Thomson CSF with Matra[12] – these are but the most prominent examples of a tissue of international connections that now link European defense firms together. Although these programs and agreements can be seen in the light of the broader trend in Europe toward a shared regional economy, as epitomized by the move to a single European market, they are significant for extending into the hitherto largely sacred area of the military technology base, and for altering from the supply side, in ways still not fully resolved, the foundation of the traditional bias toward national purchasing in the defense sector.

Dual use: a new policy for new conditions

The changes in mission, funding, and institutions for military R&D outlined above present governments with significant challenges. Few doubt that new technology remains important for military security, but the diffuse nature of the post-Cold War threat requires a different kind of technological response. Increasingly, the interesting technologies are to be found in the civil sector, requiring a new approach to managing the technology base. Changes in industry structure and the underlying trend to globalization have created a new set of actors and changed the rules of the game; international linkages are of growing importance. Governments have reacted to these developments in ways that reflect their traditional approaches and ideologies – most notably with respect to the relative weight placed on market versus government solutions – but there is evidence of convergence toward a common set of policies. In particular, in all of the countries with a history of large investments in military R&D, there is now a common enthusiasm for dual-use technology, even if, as we shall see, it means different things to different people.

Dual-use technology – that is, technology that has both civilian and military applications – potentially offers a solution to many of the problems facing military planners. It provides a means for the military to tap into leading civilian technologies and thus relieves pressure on military budgets. It multiplies the number of potential suppliers in the market and thus offsets the decline in competition caused by consolidation in the defense industry. To the extent that civilian users can benefit from technology developed by the military, it provides welcome additional justification for military spending. In countries like the United States with a political culture that rejects government intervention in the form of industrial policy, dual-use technology programs have been a way to support civilian technology behind the fig-leaf of national security rhetoric. Finally, in a world of shrinking military spending, dual-use technologies offer one possible route to defense industry diversification and economic conversion. Given this array of arguments, it is not surprising that dual-use technology has come to be seen as a panacea for the problems facing the managers of the military technology base.

The other important context for this debate is Japan, which is now the paradigmatic case for the application of the dual-use concept. In his detailed account of what he terms Japan's 'techno-nationalism', Richard Samuels (1994) shows how Japan has embedded a military production capability within a civilian economy. When, by the 1970s, the relations between technologies of civil and military origin began to change, Japan found itself in an exceptionally strong position. As the world leader in the introduction of advanced electronics into a huge variety of products and in the development of advanced manufacturing capabilities, Japan assembled an impressive commercial technology base that could be applied to military purposes, while continuing to reinforce it with work arising from military funding. To give but one example: the Japanese claim that their air-to-surface missile performs better than the US missile on which it is based because it has better gyrocompass technology. The reason, they say, is that the Japanese bearings have smoother surfaces than their US counterparts, because they were first produced for the demanding purchasers of videotape recorders. (Samuels, 1994.)

For Samuels, the fundamental lesson from the Japanese experience is that a full-spectrum commercial capability helps military production as much as do focussed defense industrial policies. But the relationship is reciprocal. Each of the pieces, up and downstream, meshes together, resulting in a diverse commercial economy that is a huge 'knowledge generator' for society as a whole.

Such thinking has now taken root in Europe, as well as in the United States. In Britain, industry and government initiatives have replaced earlier reluctance to discuss together the development of civil and military technologies. Industry is pressing strongly for a greater national effort in the development of dual-use technologies, notably through a program sponsored by the CBI on the initiative of British Aerospace, and through the Technology Foresight exercise. (Cabinet Office, 1995: Coghlan, 194: 12–13.) On the government side, the Defense Evaluation and Research Agency (DERA) is setting up new Dual Use Technology Centers.[13] The first, on Structural Materials, was launched in April 1994; a second, on Supercomputing, in January 1995, in association with British Aerospace, GEC, and Cray Research; and others have been established or are under development in the fields of marine technology, software engineering, information technology, and robotics. Related initiatives that will align the programs of the DERA more closely with those of industry have also been introduced. These include 'Pathfinder', a scheme aimed at improving the overall effectiveness of the expenditure on defense research through better alignment and exchange between government and industry. It works by inviting firms to an annual conference at which they receive a briefing on DERA plans, following which firms submit proposals for joint research. These might involve projects funded entirely from industry, but using DERA facilities; collaboratively funded projects; or entirely industry-funded ventures with a formal link to complementary work in DERA. By offering firms the opportunity for a much earlier and more proactive involvement in the MoD applied research programme, Pathfinder seeks to give industry a voice in determining what research is carried out to meet military requirements, and the chance to target its own research more effectively. It also encourages the possibility of spin-in from civil research to defense purposes.[14]

Similar developments have also been underway in France, where there was early support for dual-use technologies. For several years the government has sponsored large annual meetings to draw together defense and civil scientists (so-called 'Entretiens science et défense'). Civil agencies, such as the national space agency (CNES) are much more closely involved with defense programs than would be the case in the UK. The French outlook was clearly conveyed in an editorial in the house journal of the French procurement agency, introducing in 1991 a special issue on 'Les Technologies Duales'.[15] The editorial argued that the time was past when one could think of defense research in isolation from the civil, adding that the current economic and technological conditions required constant attention to the issue of duality in research and development. The same willingness to place defense interests firmly in the forefront of wider national technological development was further developed in the 1993 report of the Groupe de stratégie industrielle of the French Planning Commission, with its call for the creation of stronger institutional links between civil and defense interests, government and industrial, and for an appropriate division of responsibility for funding work in the field of dual use.[16]

Germany presents a contrasting case. The most powerful economy in Europe is also one of the most heavily constrained politically in terms of military equipment spending, relative to its size. Since 1985 the policy has been to draw on civilian-developed technology as far as possible; so, for example, civil satellites are used for military reconnaissance (Lock and Voss, 1997). In contrast to the United States and Britain, Germany has not maintained large state-owned military research establishments. Instead, it makes heavy use of other research institutions with primarily civil interests (such as the Deutsche Forschunganstalt für Luft und Raumfahrt, and the Fraunhofer Gesellschaft), and of private companies. These are on the whole less militarily dependent than their British or French counterparts, and in that sense much more like Japanese defense producers.

It remains to be seen how these initiatives will evolve, and whether the interest in 'duality' will amount in the end to anything more than an attempt to support military activity from the civil base, rather than to seek a more genuine integration of the two. It is evident that a certain political opportunism has sometimes attended the use of the concept. In the United States, for example, the rush to attach the label 'dual-use' to defense projects in order to qualify under the Clinton administration's Technology Reinvestment Program (TRP) backfired when the Congress elected in November 1994 turned against the TRP and cut its funding. A new Dual Use Applications Program (DUAP) was announced, but its focus is on drawing technology from the civilian sector for military use, unlike TRP, which also sought to transfer technologies of defense origin to the civil sector. (Cooper, 1996: 20.) Similarly, in France, Serfati (1995: 94) has pointed to the rhetorical freight carried by the term 'dual use', observing that in practice it tends to be applied only to technologies that can be used in the military sector. In Germany there is concern that promoting dual use will result in militarization of the civilian technology base (Attmann *et al.*, forthcoming).

Despite these rhetorical uses, however, there can be little doubt that a real change has occurred in the technological base of defense production, creating a genuine need to gain access to the civilian technology base. Information technologies in the form of communication and guidance systems now dominate the military platforms in which they are embedded, and in many cases civilian applications of these technologies are more advanced than military ones. Dual-use polities have thus targeted the information sector – for example, the programs in the United States designed to encourage government–industry cooperation in chip fabrication (Sematech), device development (VHSIC), and computer software (STARS) (Mowery and Langlois, 1996: 947–66). Similar programs have been pursued in Europe – for example, the Alvey Program in the United Kingdom, and, at the European Community level, the EUREKA and ESPRIT programs. In all of these programs, government funding, taken in part or entirely from defense budgets, has been combined with industry-supplied funds to support particular technology projects, with the emphasis on developing generic technology to be shared among the participants. The record of achievement is mixed; in the VHSIC program, for example, the contractors that survived to the later phases of the program were all established defense contractors, and the devices that they produced found only military applications (Alic *et al.*, 1992: 271).

For the military to reap the full benefits of dual-use policies they must attract new suppliers from the civilian sector: only if this happens will there be an infusion of new competition as well as new technology into military markets.[17] For society to reap the full benefits, the technology developed must find civilian applications as well as military; otherwise the policies will have succeeded only in funding military R&D under a different label. This degree of duality is not easy to achieve: military funding for research consortia and technology transfer programs is not enough. There are numerous barriers to civilian firms participating in the military market, including specialized military specifications, accounting conventions, reporting requirements, and the preference of military procurement officers for dealing with established and familiar defense contractors (who employ ex-military officers in large numbers to sustain this preference). Thus, the campaign in the United States to increase military access to civilian technology has included reforms in the procurement system to eliminate military specifications wherever possible and to reduce paperwork requirements, and similar steps have been taken elsewhere, for example, in Belgium (de Vestel, 1997). Again, the results are mixed, with implementation not yet matching the promise of the reforms. One observer has complained: 'Requests for proposals remain as thick as ever, with no significant reduction in the specifications to which bidders are expected to respond. That tells me the government can do a better job in implementing acquisition reform, because what we're seeing at the operating level doesn't reflect the work that went into the reforms themselves.'[18]

Procurement reform is essential to a successful dual-use policy. In effect, it is the means by which the boundaries between the civilian and military technology bases can be renegotiated (Gronberg, 1996: 52). If civilian companies enter the defense market in substantial numbers, they could potentially transform the military–industrial network by providing alternatives to established contractors and treating new conduits for the transfer of technology between the civilian and military sectors. They would, by definition, be diversified companies, not dependent on military business, and in that they would represent a radical departure from the canonical prime contractor committed to a 'pure play' strategy. Moreover, the move to make defense procurement more like standard commercial transactions by eliminating specialized regulations would, if successful, affect the military–industrial networks even more profoundly. It would replace the network of established relationships that rest on the expectation of repeated transactions with something more like a truly competitive market, in which transactions are not affected by the identity of the buyer or seller. Along with the undoubted benefits to government policies from tapping into the civilian technology base, however, would come the problem of maintaining the security of supply and responsiveness to government requirements that the established defense contractors have provided in exchange for their privileged position in the market.

Of course, the question remains as to whether the dual-use policies will succeed in effecting these transformations. In their complex responses to the changing environment, US defense contractors and government officials are pursuing contradictory directions. At the prime contractor level, the consolidation of the industry has, as noted above, produced a small number of defense-dependent giant corporations that, more than ever, will command government support; the alternative would be to risk losing the core of the defense industrial base. While the government's procurement reforms are aimed at broadening the market at the subcontractor and components level through entry of new firms from the commercial sector, many prime contractors have adopted a strategy of certifying a small number of suppliers in order to ensure reliability of supply. In effect, the prime contractors are choosing network over market solutions to their own principal–agent problems.

In Europe, a desire to access wider technological networks, and to push these beyond the traditional defense sector lies behind such steps as the British establishment of dual-use technology centers, even if a full explanation would have to point also to the need by the government research establishments to rationalize their own activities under the pressure of severe budgetary constraints. We also see in Europe steps toward transnational research organizations that embrace both civil and military interests. The best example is GARTEUR (Group for Aeronautical Research and Technology in Europe), which was formed in 1973 by representatives of the government departments responsible for aeronautical research in France, Germany, and the UK. The Netherlands, Spain, Italy, and Sweden joined later. Initially GARTEUR supplied an information exchange service, but it soon progressed to organizing limited cooperative programs on an à la carte basis, and then added promotion, planning, coordination, and review of joint activities. GARTEUR works through groups covering particular disciplines (e.g. aerodynamics), with representatives from government, research establishments, and industry. In parallel with GARTEUR, an industrial group was established, comprising senior R&D managers from the principal companies. It has regular meetings with the Executive Committee of GARTEUR, and has similar relations with the lower levels in the structure.[19]

Without the mechanics being entirely clear, it seems that GARTEUR provided a setting for discussions between the aeronautical research establishments of the seven member states (the 'RE7' talks) about future collaboration. It appears to be from this network that the 'Joint Position on the Future Role of the Aeronautical Research Establishments in Europe' emerged in January 1994.[20] The document envisages that: 'A federative process starting with an association, and to be followed by a partnership, should finally lead to a "Union of National Aeronautical Research

Establishments in Europe".' It proposes progressive integration of programs and facilities in the civil sector, with EU support, but notes that 'A potentially much larger field of coordinated RE [Research Establishment] involvement would be opened up by an established European cooperation in military aeronautics.' It further envisages the eventual establishment of a common defense research strategy, possibly resulting in the establishment of a European Defense Research Agency. The association came into operation in late 1994, but it remains to be seen how far and how fast the more ambitious plans will evolve. It is clear, however, that this is an example of an active policy network representing the interests of government research establishments and companies both to member governments and to the European Commission, and, with respect to the Commission, intent on lobbying for financial support for aeronautical research in Europe, regardless of the traditional civil–military boundary.

For all their apparent attractions for defense managers and governments, it should be noted that dual-use policies bring in their wake some new problems. By embedding military technology more firmly in the civilian technology sector, they open up new channels for technology transfer and thus raise problems for proliferation policy. Practices – such as secrecy and specialized procurement rules – that have maintained a separation between military and civilian technology have also served to simplify the task of controlling technology flows abroad. Throughout the Cold War, international transfers of military-related technology from western countries were managed by an international regime that combined national controls and international cooperation to deny targeted technologies and products to the Soviet bloc. Now, however, there is no longer a clearly defined enemy to target, and the blurring of the line between military and civilian technology has multiplied the potential routes for technology transfer. Policies promoting dual-use technology can only exacerbate the danger of proliferation and the difficulties in devising effective controls (Reppy, forthcoming).

Some would argue that dual-use policies cannot succeed where the underlying military strategy calls for power projection and technological dominance. In a fundamental critique of the US government's dual-use policies, Greg Bischak has argued that current dual-use technology policy is 'inherently biased toward technologies serving the power projection model of national security. In addition, the one salient economic objective of the dual-use approach, namely to make defense technologies more affordable, cannot be seriously dealt with through a process that is dedicated to forcing technological frontiers to satisfy military-strategic requirements' (Bischak, 1989: 31). In this analysis technological commonality is an illusory goal, continuously sacrificed to the demanding requirements of the military.

Thus, from the point of view of arms control, dual-use technology has ambiguous status. On the one hand, it may provide an avenue for diversification of defense firms and conversion of military R&D resources to civilian uses, thereby lessening some of the political pressures to maintain military spending in order to protect jobs. But these benefits come at the cost of earlier diffusion of military-related technology through civilian channels, creating problems for non-proliferation regimes.

Implications for governance of military R&D

The changes in the environment for military R&D that we have described present a considerable challenge for the governance task. This is perhaps ironic, because many of the changes are the direct or indirect results of government policy. Consolidation of defense industries, for example, has occurred with government consent or, in the case of France, with direct government intervention. The consequence, however, is that government policies for competition in the defense market are increasingly unsustainable, and the number of potential sources of new technology, at least at the prime contractor level, reduced. For example, in December 1996 the British firm Racal (teamed with Siemens–Plessey) and the US firm ITT, who were the only two bidders on a British army communications systems (codename Bowman), announced that they would enter into an alliance and submit a single bid in order to reduce the heavy development costs that both were incurring (May, 1996: 18).

Similarly, dual-use policies are a government solution to the problems of falling budgets for military R&D, rising costs, and the emergence of technological leadership in the civilian sector, especially in the information and communication technologies that are so critical to current military operations and strategic concepts. With the introduction of new, less dependent, firms into the military–industrial networks, however, the government must learn new ways of dealing with them: the old methods of control through practices such as military specifications and extensive government audits are not applicable to normal commercial transactions. Defense ministries cannot lock in these suppliers as they have their traditional contractors; to do so would be to recreate the problems that the turn to the civilian economy is supposed to solve. Instead, they must rely on market incentives in a market in which they are not the dominant player.

Even globalization, which is often portrayed as a *deus ex machina* on the world scene, is best understood as the product of a combination of government actions (or lack of action) and industry initiatives. Increased international collaboration between defense companies lessens the power of individual governments to control their defense industries, a process carried to the logical extreme in those cases, admittedly few in number, when foreign companies acquire national firms. There is a range of less dramatic relationships – for example, offsets for export sales or international teaming to develop new weapons – that can have much the same cumulative effect with respect to international technology flows and weakening of the national technology base. In Europe, the halting progress toward a common defense industry is already creating a regional technology base around such products as the Eurofighter 2000.

The net result of all these changes is that government-as-principals must deal with a new class of agents in order to pursue their goals for military R&D. In the military–industrial networks of the Cold War era, governments shared power with their defense contractors, but because they also shared a common set of goals, there was only limited conflict of interest. With military budgets falling and defense ministries looking to the civilian sector for new technology, shared goals can no longer be assumed. The new networks will contain actors whose main interests lie in commercial markets and who will be free to exit the defense market at will. In general, they are less likely to be part of the informal systems for sharing information, so that, even as their range of choice of agents increases, governments will have a more difficult problem with adverse selection and with monitoring progress. At the same time, prime defense contractors, fewer in number, may regain some leverage over their governments, who, if they press competition arguments too hard, could irreparably damage their national or regional technology bases.

The situation will vary across countries, reflecting differences in political cultures and, more specifically, in traditional relationships between governments and their defense contractors. Market solutions will have less appeal in France than in the UK, for example, while the combination of corporatist structures and integrated military-civilian production in Germany may provide a framework for governing military R&D that retains effective control without a lot of government bureaucracy (Brzoska *et al.*, 1996).

Conclusion

This line of analysis suggests that in the future governments will need to continue to maintain core capabilities in-house in order to assure access to the specialized technology they require, whether their concern be to maintain military capabilities against potential enemies or, as in the case of Europe, also to maintain a measure of autonomy from the United States. The means that they adopt, however, will differ. One common factor is the challenge created by globalization trends. The increased possibilities for international technology flows, whether through industrial teaming and collaboration or the increased reliance on dual-use technologies, pose a proliferation problem for international politics that will require international solutions. The governance of military R&D is, then, likely to become less and less a matter for national governments, and more a subject for regional or international solutions.

Notes

1. The Stockholm International Peace Research Institute (SIPRI) estimates that world military R&D has decreased by as much as 50 percent in real terms since the mid-1980s (*SIPRI Yearbook 1996*, Oxford: Oxford University Press, 1996: 384, fn. 6). One result of the end of the Cold War has been increased transparency with respect to military R&D, but there are still gaps in the data, especially for China.

2. For a discussion of the relationship between military strategy and military R&D, see Greg Bischak, 'The Implications of Alternative Security Doctrines and Policies for the Defense Science, Technology and Industrial Base', paper presented at the Council on Foreign Relations Study Group on Consolidation, Downsizing and Conversion in the US Military Industrial Base, New York, April 1996 (mimeo).

3. Data compiled at Cornell University show the number of cross-national teaming agreements announced each year between European defense firms increasing tenfold between the late 1970s and the early 1990s. Some of the gain must be attributed to more comprehensive reporting of such agreements in recent years, but there is little doubt that there has been a substantial increase.

4. There is a large and growing literature on the revolution in military technology. See, for example, William A. Owens, 'The Emerging US System-of-Systems', *Strategic Forum*, 63 (February 1996): 1-4; Norman C. Davis, 'An Information-based Revolution in Military

Affairs', *Strategic Review*, 24 (Winter 1996): 43–53; Eliot A. Cohen, 'A Revolution in Warfare', *Foreign Affairs*, 75 (March/April, 1996): 37–54; and Benjamin Lambeth, 'The Technology Revolution in Air Warfare', *Survival*, 39 (Spring 1997): 65–83. For a skeptical view, see Colin Gray, 'The American Revolution in Military Affairs', paper presented to the British International Studies Association annual conference, University of Durham, December 1996. On non-lethal weapons, see Nick Lewer and Steven Schofield, Non-Lethal Weapons: A Fatal Attraction? (London: Zed Books, 1997).

5. Much of this debate has taken place around the lessons of the Gulf War. For a recent report based on declassified information, see U.S. General Accounting Office, Operation Desert Storm: Evaluation of the Air Campaign, NSIAD-97-134, Washington, DC, 12 June 1997.

6. The picture is mixed for subcontractors and third tier suppliers, many of whom have successfully combined military and civilian business, see Markusen, A. (1998) The post Cold War American defense industry: options, policies and probable outcomes. In E. Inbar and B.-Z. Zilbergarb (eds), *Political and Economics of Defense Industries in a Changing World*. Portland, OR: Cass.

7. Morrocco, 1997. For further discussion of these issues, see Pierre de Vestel, Defense Markets and Industries in Europe: Time for Discussions, Chaillot Papers, no. 21, Paris: Institute for Security Studies, 1995.

8. Department of Trade and Industry/Office of Science and Technology, Forward Look of Government-funded Science, Engineering and Technology 1996 (London: HMSO, CM 3257), vol. 1, para. 4.24; vol. 2, fig. 2.3.

9. Treaty on European Union, Official Journal of the European Communities, C, 191, 29 July 1992, Article J. 4.

10. See Declaration (No. 30) on Western European Union, attached to Treaty on European Union.

11. See Keith Hayward, Towards a European Weapons Procurement Process, Chaillot Papers, no. 27, Paris: Institute for Security Studies, 1997. According to Hayward JACO has altered its acronym to JACS (Joint Armaments Cooperation Structure), perhaps in recognition of the fact that its precise legal identity remains unresolved.

12. These possibilities were thrown into confusion in December 1996, when the privatization of Thomson was suspended because of concern about the likely sale of its consumer electronics arm to Daewoo of Korea. The election of a Socialist government in summer 1997 then led to further delays in decision on the future of the defense industry, and hence also to European restructuring. David Buchan, Andrew Gowers, and Quentin Peel, 'France may split Thomson for sale', *Financial Times*, 9 December 1996: 1; David Owen, 'French defence circus ready for new season', *Financial Times*, 19 September 1997.

13. DERA is the successor agency to the Defence Research Agency (DRA), which was established in 1991 as part of a series of Thatcherite reforms aimed at bringing market-like discipline to government agencies. Under DRA the main non-nuclear defense research establishments were moved into a contractual relationship with the Ministry of Defence and run, in effect, like a business, including generating an annual return on capital. DERA remains, however, owned by MOD. See House of Lords, Select Committee on Science and Technology, Defence Research Agency, Session 1993–94, London: HMSO, 1994, HL 24.

14. House of Lords, Defence Research Agency: Ministry of Defence, Statement on the Defence Estimates 1995, London: HMSO, 1995, 75–6.

15. L'Armement: Revue de la Délégation Générale pour L'Armement, No. 29 (October 1991).

16. Commissariat Général du Plan (1993) *L'avenir des Industries Liées à la Défense, produced by the Groups de Stratégie Industrielle*. Paris: La Documentation Française, pp. 174–5.

17. There is in fact some debate about the extent to which dual-use suppliers already feature in the defense sector. See Kelley and Watkins (1995: 525–34); and Oden (1996).

18. John Kutler, president of Quartedeck Investment Partners, Inc., quoted in Anthony L.

Velocci, Jr., 'Sea Change Looming for Defense Contracting', *Aviation Week & Space Technology* 145 (19 August 1996), p. 21.

19. We are grateful to Alain Deckers for supplying information on GARTEUR. Further details come from GARTEUR Guide, published by the GARTEUR Council, 1993 edition, available from any of the participating laboratories (e.g., DERA, Farnborough, UK).

20. Of the seven 'editors' of the RE7 document, five were their nation's representatives on the GARTEUR Council; the other two were from Spain and Italy and presumably were the link persons between GARTEUR and those two countries, which do not enjoy full membership.

CHAPTER TEN

Technology, Growth and Development: The Dynamics of Catching Up, Falling Behind and Leaping Ahead[1]

Mario Cimoli and John de la Mothe

There is a crack in everything. That's how the light gets in.

Leonard Cohen

Introduction

This chapter is concerned with the political and economic thinking behind sustained development, using technological capabilities as its basis. As such, it is concerned with processes of catching up, falling behind, and, most importantly, leaping ahead. Anyone casually aware of Britain's nineteenth-century fall from empire, Argentina's lost ground as a regional, cultural and economic power early in the twentieth century, or the postwar rise of Singapore, Japan, and South Korea, would recognize the importance of this complex of dynamics. They might also recognize that traditional approaches to understanding growth and development – such as Ricardian trade theory, Vernon's obsolescence bargain, or notions of comparative advantage – do not possess sufficiently robust explanatory power.

With such developments in mind, this chapter will proceed as follows. First, it will briefly sketch political economic notions of development in a 'third world' sense. Second, it will briefly present an overview of the traditional views regarding obstacles to development. Third, it will – as other authors in this volume have done – discuss approaches that introduce 'Innovation Systems' as a conceptual framework through which economic performance across countries can be explained. The ability of having access to international networks and the difficulties of elaborating political and corporate strategies through which to achieve such access is the topic of the next section. Then a simple model is presented, while the last section attempts some modest conclusions from the perspective of technology and governance.

From DOA to DEE[2]

Paul Krugman (1999) has usefully reminded us that the term 'third world' was originally intended as a political badge of pride. 'Jawaharlal Nehru coined it to refer to those countries that maintained their independence, allying themselves neither with the West nor with the Soviet Union. But soon enough the political intent of the phrase was overwhelmed by economic reality. "Third world" rapidly came to mean backward and poor. And the term came to carry a connotation, not of righteous demand but, of hopelessness' (Krugman: 16). Today, however, the dynamic processes of globalization, complete with its science, technology, innovation and knowledge-based implications, can offer new opportunities for sustainable development paths.

The rapid advancement and distribution of information, communications and transportation technologies, the transfer of both technologies and capital from high wage to lower wage economics, the growing knowledge intensity of investment, the growing interdependence of nations, and the subsequent erosion of sovereignties have all rendered the socialist dream of isolation moot in light of the global quest for economic growth and development. It has also been rendered archaic given the widespread policy search for generating economic opportunities, capturing an increasing world share of higher value-added activities, and distributing benefits across economies and down into the more marginalized elements of a society. Once protected elites became subjected to the tsunami of globalization, the challenge to science, technology and governance became unavoidable. By focussing on knowledge, education, skills, technologies and innovation, public policies – as being conceived and deployed in nations as diverse as South Korea, Taiwan and Singapore – are beginning to give rise to terms such as Newly Industrializing Countries (NICs) and/or Dynamic Asian Economies (DEA) thus suggesting that trend is not destiny, that countries may not be relegated to a slow or eroding development curve and instead that they may be able to move rapidly up the value-added curve.

In the industrializing world, two different patterns can be distinguished. One is the DEA pattern, where the modes of governance of a large variety of exports-led incentives and institutional systems supported the process of upgrading technological capabilities. As it as depicted, for example, we can see the acquired competencies for the design of new products in high technological sectors. The second regards the largest Latin American economies. Around the mid-1980s, the set of rules under which the Latin American economies operated changed dramatically, particularly in the cases of Argentina, Brazil, Chile and Mexico. The new set of rules included a different trade regime of incentives determined by the elimination of many of the barriers in effect during the previous decades; in addition, the government adopted a more limited role in the development process which implied a modification in the management of the nation's macro-economic variables. This was an institutional change that created a whole new environment in which the firms and institutions operate, and there are important implications in their interactive interplay in developing technological capabilities. However, after all, the above-mentioned globalization process of Latin American economies has not yet further developed in a pattern based on dynamic comparative advantages and upgrades of technological capabilities. In general, the specialization pattern that dominate most of Latin American economies is based mainly on static comparative advantages, abundant natural resources and 'cheap' labor forces.

Broad traditional obstacles to development

As Nehru's optimistic political framing lost currency, the transition reflected two broad meanings. At an early stage, a dominant approach to development and the economics of growth deferred to a neo-Marxian set of presumptions – typified by the classic contribution of Paul Baran (1956)[3] – regarding growth and accumulation and the disruptive effects of technological change. Ostensibly two features are common to all traditional developing societies. One is low per-capita income. The second is the absence of growth.

Economically, the use of terms like 'third world' have tended to obscure the diversity of specific conditions that detracted from potentials being realized. In Asia, Latin America, Central America and Africa the range of political and cultural experience is vast indeed. Some elements of economic 'backwardness' can be traced to colonialism and civil war.

Political obstacles to development traditionally include the lack of political stability and the presence of political independence. 'Bad' economic policies can also prevent economic growth from getting underway. And they can bring growth to a halt once it has started, so countries can fall behind just as they can progress.

On the basis of development economic theory, it is well-known that trade, free trade and the internationalization of production were not necessarily neutral in the different countries' growth paths (de la Mothe and Ducharme, 1990; Dosi *et al.*, 1990). Thus, some countries have selected a road that is characterized by increasing gaps or immiserizing growth. Trade liberalization and foreign investment flows are not the only elements that help to create a prosperous development path. The firm's and sectoral learning patterns as well as overall national capabilities are dynamically coupled via input–output flows, knowledge-spillover, backward and forward linkages, complementarities – and context-specific externalities. Together, they contribute to shaping the organizational and technological context within which each economic activity takes place. In a sense, they set the opportunities and constraints that each individual production and innovation process faces – including the availability of complementary skills, information on intermediate inputs and capital goods, and demand stimuli to improve particular products. This has a direct link with the analyses that focusses on structural change and development (here, within a vast literature, contributions that come immediately to mind range from Hirschman to Rosenstein Rodan, Gerschenkron, Chenery and Sirquin, among others.)

A traditional statement of that rationale is stated by Hirschman as follows:

> One suggestion along this line was that development is accelerated through investment in projects and industries with strong forward or backward linkage effects. I argued that entrepreneurial decision making in both the private and public sectors is not uniquely determined by the pull of incomes and demand, but is responsive to special push factors, such as the linkages, emanating from the product tide. By now, the various linkages and their interaction have taken on a new character and importance. They appear to constitute a structure that is capable of generating an alternative path toward development or underdevelopment for the different staple exporters. In other words, some of the principal features of a country's development in the period of export-led growth can be described in terms of linkages deriving from its staple. (Hirschman, 1977: 70 and 80)

Taken together, these typical macro views on impediments to development and growth also embody the view that the development process invariably insists that economies pass through the historical experience of the advanced industrial economies. Thus the history of the consumption intensive industrial revolution and of classical political economics became the focus and the template for many development economists (Hirschman, 1971; Griffin and Gurley, 1985).

International obstacles to development and growth can also be found to traditional views regarding gains from trade. For example, the theory of comparative advantage states that nations with differing endowments of capital, labor and natural resources will gain by specializing in those areas where their relative costs of production are low and importing in those areas where their relative costs are high. Furthermore, the greater the differences in endowments between countries – and the differences between rich and poor countries are indeed great – then the greater the gains from trade will likely be. However, in order to participate in a high value-added sense and not be simply a source of low wage labor and production for advanced economic units, then a country must have a capacity to absorb and retain talent, to produce new knowledge, and so on.

Technology, trade and development

The concept of 'takeoff' is frequently cited in the development literature and is in many ways at the center of Walt Rostow's (1971) analysis of the stages of economic growth.

Since the 1960s, the central purpose of most contributions in the field of technology and trade has been to highlight the crucial importance of technological change and innovation in explaining the international trade pattern (e.g., Freeman, 1963; Hirsch, 1965; Hufbauer, 1966; Posner, 1961; Vernon, 1966). This approach has stressed international *asymmetries in technology* as the main determinant of trade flows and specialization patterns.

In Posner (1961), the pattern of trade is explained by the initial asymmetric access to technological knowledge in a world characterized by similarities in demand patterns. In this context, trade between countries will be maintained if the differences in national abilities to innovate and imitate persist. After a time lapse, most countries can imitate the new commodity and restore technological parity, also eliminating the basis for trade. Freeman (1963) and Hufbauer (1966) have stressed the differences in the factors which determine specialization before and after the imitation process takes place. Thus, during the innovation process the effects of patents, commercial secrecy, static and dynamic economies of scale prevail. However, once imitation occurs, specialization will be determined by the traditional process of adjustment in production cost and competitiveness.

In Hirsch (1965) and Vernon (1966), technological asymmetries are associated to distinct phases in the evolution of a technology and a specific international distribution of innovative capabilities in the production of new commodities. Innovative advantage is the main feature of the initial phase, explaining the production of new commodities in the advanced countries. Over time, technology evolves into a mature phase, characterized by the standardization of products and processes. In this latter phase, international competition is based on production cost advantages and technology can be transferred to less developed economies, whose comparative advantage lies in their lower real wages. In this respect, the pattern of trade is considered a process of technological divergence and convergence, for which the innovative process induces divergence while imitation and diffusion induce convergence between countries.

In doing so, many of these studies have undoubtedly scored points with policy-makers who have increasingly come to recognize the significance of technology for international competitiveness. The basis for policy of these contributions remains poor, however. This is in fact not surprising. The introduction of 'technology' in any kind of model, whether of the classical or neo-classical sort, raises many challenges. The complexity of the phenomenon of technological change on the one hand (with its dual impact on efficiency *and* new demand) and the essential dynamic 'change' perspective implicit in the concept of technological change on the other, are difficult to handle in their normative proposal in any normative economic model.

Recent formal approaches put increasing attention to uneven international technological change as an engine of growth.[4] These approaches have pinpointed the importance of the interplay between absolute and comparative advantages as determinants of the participation of each country in world trade, the dominance of technological gaps in the process of international specialization, and the bounds imposed by the dynamics of innovation and trade on the 'growth possibility sets' of each economy. On the determinants of absolute and comparative advantages, technological gaps – in term of product and process innovation – and institutional asymmetries – in terms of the main form of organization of labor markets – contribute to determining the pattern of specialization and its evolution overtime. On the demand side, on the other hand, asymmetries in national consumption patterns, regarding price and income elasticities, play a crucial role on the interplay between specialization and macro-economic level of activity. Finally, trade balance conditions determine the growth rate differential of trading economies.

Under such an approach, it has been demonstrated that the growth of relative trading partners depends not only on the demand structure of each economy constrained by the conditions of the balance of payments, but also, particularly, on differences in technology. Furthermore, the technological gap is introduced as one of the main variables explaining the pattern of growth possibilities through the effect of what we will refer to here as the technological gap multiplier which accounts for the initial distance between countries in terms of technological capabilities (Cimoli and Soete, 1992).

Wide differences apply to the capabilities of developing new products and to different time lags in producing them after they have been introduced into the world economy. Indeed, the international distribution of technological capabilities regarding new products is at least as uneven as that regarding production processes. For example if one takes international patents or the number of discrete innovations as a proxy for innovativeness, the evidence suggests that the club of the innovators has been restricted over the whole past century to a dozen developed countries with only one major new entry, Japan. Second, the processes of development and industrialization are strictly linked to the inter- and intra-national diffusion of 'superior' techniques. In a related fashion, as already mentioned, at any point in time, there is likely to be only one or, at most, very few 'best practice' techniques of production which correspond to the technological frontier. In the case of developing economies, the process of industrialization is thus closely linked to the borrowing, imitation and adaptation of established technologies from more advanced economies (Archibugi and Pianta, 1992; Kim, 1997a). These processes of adoption and adaptation of technologies, in turn, are influenced by the specific capabilities of each economy.

National innovation systems, capabilities and economic performances

A significant body of recent literature exists explaining the importance of institutions and their role in economic and industrial development. In particular, regarding the Pacific Rim newly industrialized economies (NIEs), the works by Amsden (1989), Cantwell (1991), de la Mothe and Paquet (1998a) and many others help to understand how not only institutional successes are achieved, but also institutional failures as well. Bardhan (1996) analyzes such issues as one of coordination, which has to be seen in terms of the interaction of distributive conflicts with state capacity and governance structures. The author suggests that the success of institutions in some NIEs (namely South Korea and Taiwan) has to be understood in terms of the capacity of establishing and applying rules of performance criteria, so that, for example, credit allocation by the state was tightly bound up with export performance. In this way, international competition was used to foster internal learning. Most of these approaches point out that growth is not automatic. Growth needs a 'social capability' which can be viewed as a 'rubric that covers countries' levels of general education and technical competence, the commercial, industrial and financial institutions that bear on their ability to finance and operate modern, large-scale business, and the political and social characteristics that influence the risks, the incentives and the personal rewards of economic activity including those rewards in social esteem that go beyond money and wealth' (Abramovitz, 1994).

These patterns provide examples of the functioning of what could be understood as a National Innovation System (NIS).[5] In a narrow way, we shall propose a concept and representation of NIS whereby its main feature will be related to empirically capture the main features that explain technological asymmetries and their interplay with growth performance. It now proposes the idea of a set of technological capabilities that essentially refers to a firm's, organization's and country's ability to solve both technical and organizational problems, on the one side, and performance – as measured by such variables as competitiveness and growth – on the other. In between these two, and shaping their interaction (and therefore causing the magnitude of the span that exists between the two) lies the innovation system, acting at both the national and regional levels and therefore possessing an inherently local nature.[6]

At a national level, innovative capabilities are clearly difficult to quantify. Thus, a country's efforts to create these could be considered a proxy for the main capabilities available in each economy. Capabilities can be identified as those which pertain to the following groups:

- educational (literacy rate, secondary and tertiary level enrolment ratios, third level students in math, science and engineering);

- R&D efforts (scientists and engineers in R&D, R&D in GNP, ratio of private and public R&D);
- technology transfer efforts (FDI stock, imports of capital goods).

An essential aspect of capabilities regards their interplay with the catch-up argument. In particular, when a country that is behind the world innovation frontier is considered, most of the capabilities variables could be considered as a support for the scope for imitation. Thus, through the approaches introduced above, a catch-up process based on borrowing, imitation and adaptation of established technologies from more advanced countries is captured (Abramovitz, 1989; Fageberg, 1995; Gomulka, 1971; Maddison, 1991).

Economic performance depends on how each country implements policies (to stabilize the macro-economic setting and to promote the industrial sectors) and organizes its institutional infrastructure and efforts. The relevant variables which can be considered as a proxy for the scope of catch-up are: GDP per capita, rate of growth of GDP, exports as percentage of GDP, productivity growth, technological sectoral distribution for GDP and exports in terms of technological classification (e.g. Pavitt's taxonomy), international market share of exports and imports and employment in manufacturing sectors.

In this framework, the possibility of institutional failures becomes incorporated into a broad structure which is able to account for the interactions among the main agents in the process of development. The essential feature of this system is constituted by the interface between competencies, efforts and performance and the role that the NIS plays in it as the wider representative of institutions (both public and private). Knowledge flows are embodied in individuals and their organizations, as already stated, and therefore it is obvious that the central part in the system is played by a collection of institutions. Most countries are characterized by different types of institutions which, however, could be distinguished according to how these contribute to the development of technologies and the role that they play in the whole system. Thus, one can think of the following institutions as a sort of representative sample of those actually existing in different countries: higher education institutions (HEIs), research and technological development organizations (RTOs), industrial research laboratories (IRLs), government research institutes (GRIs), military research institutes (MRIs), agencies for education and training programmes (AETs), certificated research and technology organizations (CRTOs), technological information centers (TICs), organizations supporting science parks (OSSPs), and technology transfer institutions (TTIs).

A few examples of how the system functions are provided by the historical case studies discussed by Dosi *et al.* (1990), showing the mechanisms through which the NIS, by fostering R&D, has been enhancing the capabilities of firms and industries, ultimately inducing an improvement in performance. The same type of relationship is also confirmed by the case studies recently conducted by OECD that revealed the extent and types of collaborations between enterprises and the public sector research base (formal collaborations – such as commissioned research, joint R&D projects, co-patenting and co-publications; informal transaction – informal contacts and use of published scientific knowledge; spinoffs from universities, and transfer of technology to enterprises – patents and product developments).

At a more specific level, and in particular by focussing on the educational policy, another example of the relationship running through the NIS to enhanced competence and better performance is provided by the analysis of the Taiwanese experience supplied by Nelson (1993) and Della Giusta (1996).

In Germany, Japan and Sweden it has been often observed that, at least in most of the postwar period, education and training systems have been particularly efficient in providing people with the requested standards in terms of knowledge and skills. University and government laboratories can therefore be considered as an important source for technological capabilities for the firms. In this respect, one can also mention the cases of Korea and Taiwan, emerging countries where it seems that education-led growth contributed to shift international specialization from standard products to much more sophisticated ones, gradually characterized by a higher content of technological complexity.

In other words, a specialization increasingly oriented toward innovative commodities and practices has been often supported and associated, particularly with the successful cases, by long-term planning initiatives and public investments in 'human capital'; while – as counterfactual examples in the cases of Argentina and Brazil – it seems that the weak links between the science and education systems and the industrial and R&D system contribute to an explanation of their weak competitive performances. Taking into account this institutional supportive environment reduces, at least in part, the serious analytical and empirical difficulties that the scholars have faced in explaining within an incrementalist perspective the incredibly fast technological learning trajectories experienced by some Far Eastern firms (Amsden, 1989; Canuto, 1994; Freeman, 1987; Hobday, 1995).

Moreover, taking this analysis to its logical extreme, one might say that the international performance can be obtained only according to the differences of NIS capabilities displayed in each economy. Thus, international competition could be based on a 'protected piece of the NIS capabilities', or on industries supported by government policies in terms of resources for science and technology development. Such has been the case for Korean *chaebols* as well as for the American computer, software and semiconductor industries, supported by research resources from the Department of Defense (Nelson 1993).

A more recent analysis referring specifically to the NIS approach in the NIEs can be found in the works by Katz (2000), Kim (1997a) and Lall (1997). The argument here puts the emphasis on the 'message' that growth and catch-up potentiality are clearly related to country historical paths and development of a local NIS. Institutions, industrial and S&T policies – as those that support science, human capital and learning capabilities – are the main variables introduced to explain differences between the NIS developed in Latin American and Asian NIEs.

An implication of this view is related to a broader set of approaches that look for a framework in which the mechanisms that support technical change and innovation could be understood, so that governments could form and implement policies in order to influence the processes of innovation. Moreover, by systematizing the difference between competence and performance, it could also become possible to create a concept of measurement of the 'goodness' or 'badness' of the NIS. Through the representation introduced above it is possible to explain why technological gaps among countries reproduce themselves over time due to the fact that individual behaviors (in response to the existing patterns of incentives and opportunities) produce sub-optimal collective outcomes. In other words, the existence of diverse institutions and organizations, and their modes of interaction, determines specific national systems of innovation which over time present certain invariant characteristics which account for their phases of relative 'technological success and failure'. When organized appropriately, an NIS is a powerful engine of progress. Poorly organized and connected an NIS may seriously inhibit the process of innovation (Metcalfe, 1995).[7]

Capabilities of accessing international networks

The purpose of the above sections has been to discuss the empirical and analytical features that support an NIS analysis in order to understand their impacts on the sources of technological asymmetries and growth. However, most of these studies have implicitly stressed that growth benefits can be mainly obtained from local and domestic efforts. Today, policies oriented to increase local investment in technological variables and linkages – that essentially refer to the improvement of local functionality of the system – are not enough. In fact, with the growing internationalization of technology and production, the improvement of capabilities has to be related to the ability of accessing international networks where knowledge and technology are produced.

Since the world economy is becoming more interdependent, this phenomena is adding a new dimension into the analysis of the technological asymmetries. The growing interdependence between all the economies and regions, through raising commercial and financial flows, along with domestic institutional constraints, are affecting the traditional trail of knowledge elaboration and technology.

In fact, today, firms and countries have large opportunities of accessing knowledge and technology at the international level. The globalization of industrial research – traditionally a headquarters function – is also increasing in the 1990s and this will continue into the twenty-first century. The factors driving this are becoming more complicated and diverse as firms join university-based and private–public research consortia, cross-industry strategic alliances and so on. Traditionally, the industrial propensity to invest in research abroad was highly correlated with the internationalization of the firms functions, such as production and marketing. In an economic sense, *market access* is still an important factor in the globalization of research, particularly in those sectors that have high levels of foreign manufacturing (de la Mothe and Ducharme, 1990). As product cycles shorten and technology becomes increasingly complex and specific, more R&D facilities are created near foreign plants but also, shared R&D activities are developed to reduce the time of transfer of knowledge. In recent years, *personnel access* and *knowledge access* have become the more significant factors in the globalization of research. Research has shown, for example (Freedman 1999), that personal contacts and proximity are a *sine qua non* to network realization and research productivity. Cyber-networks, without loyalty, trust and interpersonal relations, are not sustainable (de la Mothe and Paquet, 1999).

A new international scenario: policy, globalization and complexity

Policies for international accessing in an economic environment in which firm-level profitability and state-level competition is no longer grounded in tangible goods and natural resources (which themselves feature quantity-based pricing), economic agents now have to trade with intangibles, which are much more difficult to monitor, price, and produce. This results in an environment of economic competition which is not only more fierce and more open to competitors, but is also open to greater risk. This is true for two reasons. Globalization means that firms must compete at great distances from their home. Production abroad becomes costly. Being represented and distributed in far off markets is costly. And servicing clients in foreign countries in a timely fashion is costly. Together these mean that firms must stretch their resources considerably.

Technology is a co-aspect of globalization which drives up risk in so far as the most advanced technologies emerge from research and development (R&D). R&D is based on creativity and discovery. Thus, R&D can be considered inherently risky as it cannot be guaranteed to generate commercially viable or useful results every time. Nor can the next breakthrough in any given research field be guaranteed to emerge from labs in this-or-that location. As has been noted, in order to overcome the twin dilemmas of both global presence and technological advance, organizations in regional clutters and networks increasingly operate through cooperative strategies. In short, they are cooperating in a complex environment in order to capture value from ideas, innovation and people.

The new complexity has created an environment of uncertainty because the velocity and specific intensity of technical innovation (Figure 10.1). Complexity in such a context can be expressed in terms of the systematic overlap of economies and technological factors. Some of them can be now briefly summarized. Increased costs for firms that are associated with the globalization of production and markets, as well as by the costs of carrying out research and development. Increased risks associated at the firm level by sunk costs of in-house R&D activities, by decreasing product life cycles and, at the more national level, by both the increased pace of competition and the increased emergence of international interdependencies which are restricting, to some degree, the flexibility and autonomy that governments may feel when dealing with such matters as financial markets and market access. The increased pace of competition, fueled in part by the availability of worldwide telecommunication networks and information systems, and coupled with the concomitant reduction of reaction times when faced with market signals, emergency response, or environmental emergencies, are a good examples of the nature of the current world complexity.

Local capabilities and international accessing: a need for governance coordination

Let us refer to the case of the modes of governance technology and development in Latin American countries. In an extreme synthesis, from the afore-mentioned points, it is clear that access to international networks is important because they are the

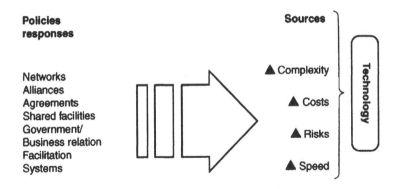

Figure 10.1 Summarizing complexity

expression of cooperative exchanges between economic agents, and generate effects both at the micro-, meso- and macro-economic levels. The importance of networks does not rely only on the support of information inflows, but in the specialization of knowledge as well. While production activities become more complex and economic agents more interdependent, complementary actions among firms and organizations are necessary in order to improve the global market position of the former. However, equal distribution of benefits can not be generalized when interdependencies and networks are developed across countries. Benefits of networking activities are not necessarily equally distributed. In an interdependent world, who capture the benefits of the above networks and externalities is an important element which should be more frequently evaluated. For example, the modernized Latin American firms are becoming increasingly globalized in terms of the orientation of production and capacity to acquire foreign technology. More important, the liberalization process provides incentives to develop networks with firms and institutions located abroad (Cimoli, 2000).

What is the main consideration related to the effects of innovation in the experiences of some Latin American countries? We know that the building of national innovation systems is a social process that involves knowledge and learning, not only about the way to do things, but also essentially of how to produce cooperative environments in which the economic performance can be intensified. Consequently, the modes of governance of local institutional frameworks, educational and training systems, R&D incentives and industrial relations affect the innovation system itself and the kind of knowledge networks which support the innovation system. Our conjecture here is that Latin America presents a rich case where the modes of governance have been exclusively 'delegated' to the market, driven by the processes of trade liberalization and globalization of these economies. Consider as an illustration the following stylized facts.

(i) Most of the Latin American production structure has gone through important changes from the import substitution period through the new scenario characterized by macro-economic stabilization policies. The largest economies, after trade reforms, have increased their participation in the world arena in terms of exports and imports. Most of the surviving and efficient firms (multinational corporations and large domestic firms) have increased their export of final commodities and automobile components, chemicals, plastic products, glass, beer, electronics, steel, cement, etc. On the other hand, their imports of intermediates, machinery and equipment also have grown. Thus, the image that we have is that countries such as Mexico and the Central American nations have greatly globalized their production activities, and a new pattern of specialization in the global production chain is emerging.

(ii) In recent years, a new pattern of sectoral and production chain has appeared, while at the same time economic activities are coordinated and integrated across geographic borders. In fact, production activities now are widely recognized to be the center of gravity for the newly industrialized economies' increased participation in the new 'world specialization'. This new globalized scenario has increasingly modified the nations' competitiveness, as well as their integration in terms of production capacities across firms, industries and nations. Industries and firms are now integrated in an international network according to different types of linkages designed as networks of contractors, or they have formed more coordinated, integrated and organized production chains in different sites around the world. Many other factors give specific shape to these networks; one such factor involves the rules of action for NMEs as they decide whether to invest locally or export, based on the trade-off concerning proximity of trade for local investment compared to the gains in concentration or scale that come with exporting.

With a difference to timing, it is possible to infer that an 'optimal cycle' of globalization, which relates growing production capacities to dynamic and higher technological efforts, will prevail in the long term. At the beginning of this process, foreign firms start their production activities in a domestic location, adapting organization equipment around their main headquarters-based business. Manufacturing activities are located mainly on the basis of labor costs and local markets that have specific regulatory frameworks (think of NAFTA as introducing a new regulatory framework). For some authors, local technological competencies become an early ingredient in persuading multinational enterprises (MNEs) to establish subsidiaries in a large range of industries (Cimoli, 2000). However, in general, in the following stages, local firms are expected to develop their own technological competencies by adapting the production organization/processes and product designs. In the last stage, R&D is expected to move ahead, establishing research centers and linkages with local centers that specialize in the production of applied research and the formation of human capital.

This last phase, associated with the new phenomenon in the globalization of industrial research activities, generally has occurred throughout the most industrialized economies in the OECD area, as more of the multinationals' investment is directed toward research abroad and the acquisition of laboratories. Nevertheless, recent evidence on the large multinational firms' contribution to the world's technology shows that, in spite of increasing talk about the globalization of the technological activities of large firms, they remained remarkably oriented toward domestic activities. MNEs prefer to keep technological activities at home, in contrast to their attitude toward production activities. For foreign firms, the overall evidence seems to point to the contrary of the expected convergence. Technical literature cites international evidence about large MNEs keeping most of their technological activity at home (Patel and Pavitt, 1995: 37). In fact, analysis carried out through empirical studies at the meso- and micro-industry level in countries such as

Argentina, Brazil and Mexico has shown that this hypothetical phase remains a 'dream' for most of these countries.

Under these circumstances, foreign direct investment (FDI) refers to the activities and decisions that MNEs make. These activities and decisions, developed considering international production, exert a strong influence on the direction of trade flows, scale and content as well as on trade specialization, competitiveness and the host and home countries foreign trade balances. This is the case of host countries in Latin America. In fact, to a significant degree, their patterns of trade specialization and performance (for example, international competitiveness) then can be analyzed as the outcome of the processes that are a result of the MNE's decisions concerning the location and quality of direct foreign investment. In this context, regional integration through NAFTA has played a crucial role as an institutional regime or framework that has supported incentives for the MNEs. Today, technological development occurs mainly in the MNEs' home bases, and only a small portion is transferred to host countries. This process determines, on the one hand, that Latin American countries do actively participate in the globalization of production and, on the other hand, that their participation in the globalization of scientific and technological activities is very poor. As companies transfer only some of their R&D activities, we can expect that the present concentration of corporate R&D will by and large lead to an even stronger international divergence of technological development. The internationalization of R&D is carried out within developed economies and regions with already-proven technological advantages. Technological cooperation between firms seems, in practice, to exclude firms that do not already have an established reputation within the developed economies. This view supports the results obtained by empirical research on the organization of research activities in multinational firms; here, it is clear that even multinational companies perform most of their innovative activities in their home country (Patel and Pavitt, 1995) and, with some specificities (Cantwell, 1989; Cimoli, 2000).

(iii) Most production activities in Latin American countries have increased their demand for the knowledge and technology that foreign sources provide. Our evidence indicates that most countries have modernized their plants that specialize in exports, which suggests that industrial adjustment has occurred preferentially through process innovation such as the improvement of production organization, improvement of skills and adaptation of machinery and equipment – not the renewing of fixed capital – which would permit the MNEs and large domestic groups to achieve a better competitive performance. Moreover, there are three main reasons why the dynamics of inter-industry flows simply is not functioning to improve R&D efforts and linkages with the local institutional framework. For example, in Mexico and Central America, Maquiladora operations dominate the production of science-based components, thus allowing for very limited links and flows to other domestic suppliers of intermediate goods. Specifically, when analysis is developed for the most recent years, it seems to confirm that the Maquila industry is one of the leading actors in the industrial modernization. The diffusion of this type of industry introduces only very weak connections with the domestic productive firms and institutions. The 'maquila innovation system' mainly supports and stimulates networking activities in the firms and institutions located abroad, thus reinforcing the knowledge and technological advantages in the developed economies.

(iv) The imported equipment used throughout the industrial system replaces the learning capability that could accumulate in specialized domestic suppliers of equipment in a well integrated industrial system. The main changes could be observed in the modes of how sectors and the type of firms (considering foreign firms and non-foreign firms) are interlinked with foreign production networks and sources of technology. Specifically, the pattern related to R&D efforts and other modes of technology transfer has been replaced mainly by greater integration with imported inputs, stronger linkages with foreign engineering services and institutions (such as universities and other research institutes) for the most successful export sectors. Their direct contribution to R&D and technology transfer is not substantial. The evidence on R&D activities and technical collaboration (technology transfers) shows that the efforts and local interactions for these kind of activities are scarce and scattered (Cimoli, 2000; Katz, 2000).

(v) The personnel employed in activities such as R&D, quality control and local adaptation of design mainly interact within the multinational firms where they work and, furthermore, those firms are characterized by reduced linkages with the domestic higher education institutions, local research centers and laboratories. In this context, for example, universities show increasing efforts to improve and create linkages with the production system. However, those efforts are inhibited by two principal factors: first, the bureaucratic organization in most public universities, and second, the more modernized industrial sector and the science-based sector's demand for knowledge from institutions and research centers abroad.

The above-mentioned factors have inhibited local networking activities. Moreover, local networking activities did not have sufficient support, in terms of linkages between the different agents in most of the innovation systems in the Latin American countries. In other words, the production system has modernized a small part of the economy, due to the effects derived from the opening-up of the economy; however, this process has not been accompanied by an increased effort to stimulate the creation of local networks, such as a non-market system of linkages, a business culture and institutions that enable firms to interact with each other.

Most policy-makers have held that the liberalization process is a sufficient condition and that they only need to support the acquisition of foreign technology and to locally capture and absorb the benefits from the internationalization of trade, investment and technology flows. In fact, policy-makers point out that the openness of the economy has expanded the nation's technological opportunities, thus improving its technological efforts. Up to now, the modes of governance that are likely to be specific to local institutions, local regulatory framework and 'authorities' for privatization rules has only played a passive role. Furthermore, efforts to promote more integrated systems, which would allow firms to rapidly address technological and organizational problems, have not been carried out.

In general, our conjecture is that the outcomes of this process at the micro-economic level are that the technological trajectories of each economy can be completely divergent. The instrumentation of networks generates conditions for gaps in the creation and absorption of knowledge and of technical change. In this sense, the macro convergence of economies does not mean equality of the members, yet implies different capabilities in capturing the benefits of networking activities. The micro-economic divergences observed in the technical paths of different countries at the regional level is explained by the differences in the institutional framework of each economy, i.e., the functioning of networks does not imply uniformity of the technical procedures.

In this context, it is possible to infer that the technological performance of the economies is mainly determined by the *composition, size, flexibility* and *international accessibility* of the National Innovation System. On one side, it is possible to find economies that are successful in developing local abilities through strength of linkages between different parts of the NIS. While in the other side, it is possible to find prosperous economies with intensive connections with the international technological knowledge system. Fundamentally, there exists an irrefutable fact – the economic success of a society relies upon the abilities they develop to generate and incorporate knowledge and techniques produced in other economies.

Forging ahead, catching-up and falling behind: a simple model

By referring to the trade and growth analysis introduced above, it is possible to draw an interpretation and representation of the main interplay between the NIS, capabilities and growth performances.

Within the learning economics, the mechanisms of adoption and learning substantially modify – and add new interpretations to – the cost functions faced by the sectoral country productivity. The description of the process proposed here explains how unit costs decrease in accordance to a technologically determined learning curve, with capabilities clearly possessing a cumulative character. This process has been adapted from the evolutionary-structuralist models which introduced some sort of Verdoorn-Kaldor law for the explanation of the interplay between learning, dynamics of productivity and trade specialization. The introduction of endogenous dynamics increasing returns is displayed by a mechanism which explains gains in specialization in terms of an increase in exports and higher international market shares.

In figure 10.2, we have a representation of the interplay that exists between productivity (π) and exports (z),[8] whereas each schedule π-z shows that changes in productivity are leaded by exports increasing. Thus, under the assumptions that a change in productivity affects the whole range of commodities produced,[9] an increase in the country's exports – higher participation in the world economy and market shares for the commodities already exported – will lead to a higher productivity via learning mechanisms, organizational change and quality improvements.[10]

By referring to these relationships, different schedules of specialization and productivity can be obtained. Looking at Figure 10.2, one can discern an increase of domestic capabilities (literacy rate, secondary and tertiary levels enrolment ratios, third level students in math, science and engineering, scientists and engineers in R&D, R&D in GDP (private and public R&D), access to international knowledge networks, direct FDI stocks, imports of capital goods, etc.) moves the function on the right side and will increase the number and type of commodities exported. In the high NIS capabilities area, the number of commodities produced and exported will be increased (new international markets could be gained for 'new' commodities that have been already imported or produced only for the local market). Two simple and

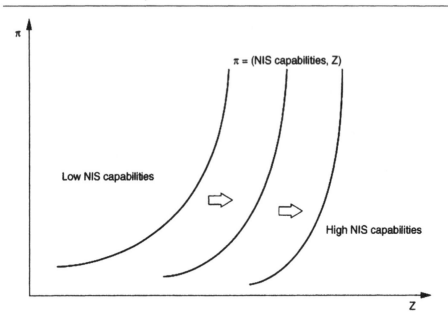

Figure 10.2 NIS Capabilities and specialization

schematic processes are described here. First, a country can gain in terms of dynamic comparative advantages when a path of productivity growth is taking place. General support could be found, among others, in the learning mechanisms, organization of production and product quality improvements. Second, a country's efforts to increase its participation with 'new' commodities in the world market are mainly embodied in the changes of NIS-capabilities addressed to support innovation, its diffusion and access to international knowledge networks. In sum, here we describe the two patterns with respect to the DEA and Latin American economies. At least, the pattern in Latin America can be described by the movement along a single schedule, thus these economies improve their productivity on the basis of a modernization process held in the larger exporter firms and their comparative static advantages. The DEA pattern, on the other side, seems to be characterized by an increase of their productivity, along a single schedule, and the movement of the schedule on the right side. In other words, these are economies that not only gain in terms of a continuous process of modernization but they have created new capabilities and competencies for the design of new products and process.

Figure 10.3 graphically represents the interplay between NIS capabilities and catching-up processes. Looking at the left side, think of a 'NIS capabilities frontier' which could be identified as the actual created asset that each economy has shown to have built up in order to support innovation and its diffusion. Each π-y schedule is built up on given NIS capabilities. The Figure indicates that an increase of domestic productivity will lead to a higher domestic income (with respect the world economy). In other words, an export-led growth approach is assumed here.

On the right side of the Figure 10.3, the growth trajectories are displayed (whereas y approximates the GDP as a proxy for the scope for catch-up). The export leading growth process is taking place on a single schedule relating y changes over time.

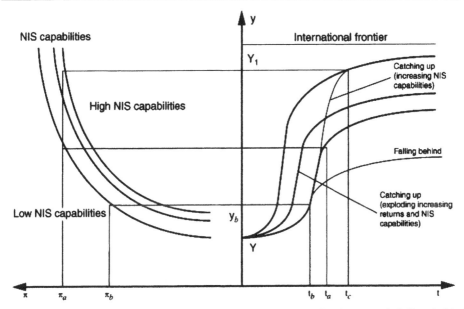

Figure 10.3 NIS Capabilities and trade growth performances: forging-ahead, falling behind and catching-up

Thus, if our country is behind the international frontier, a process of catch-up is taking place when its productivity increases and learning mechanisms (associated with organizational changes, improved adaptation of imported capital goods and quality improvements, etc.) in the production activities are obtained.[11] Two limits are represented when the catching-up path takes place $(Y)_1$ and $(Y)_2$ which could be considered as physiological bottom-up of relative growth possibilities; in particular, y_1 represents the 'international frontier' capturing the performance attained by the developed economies.

Again, on the left side, a successful effort to increase this NIS capabilities asset would shift upward the schedule π-y. The same could be obtained when the set of institutions, which jointly and individually contribute to the development and diffusion of new technologies, improves the transfer and absorption capacity of knowledge flow and, consequently, the diffusion of the innovation process and the access to international knowledge networks. Thus, an effort to increase NIS capabilities and improve its functionality would produce a jump of the schedule y-t reducing the gap from the international frontier.

In other words, a stronger effort to obtain a functional (well organized) NIS and higher capabilities of accessing into international knowledge networks will lead to an upward shift of the schedules π-y and y (the income increase from y_a to y_c). Thus, a higher participation and higher income with respect to the world economy could be reached; and, a process of catch-up is taking place and reducing the growth gap from the international frontier.

Most of the effects described earlier can be neutralized by a higher effort to increase the NIS capabilities and improve its organization in the advanced countries. In this case, a jump of the international frontier (y_1) and an increasing gap from the others' growth trajectories could be obtained. A scenario of a stable gap in the growth trajectories in terms of the international frontier could be viewed as the result of similar efforts in NSI capabilities.

The catch-up process introduced here places emphasis on differences in the scope for export-led growth and a country's potential to support innovation systems and international access.[12] In this sense, two are the mechanisms introduced and described in Figure 10.3. First, a country could exploit growing productivity on the basis of improved competitiveness and increased exports. This process is building-up on a given NIS capabilities framework. Thus, learning mechanisms, organizational changes and quality improvements in the production activities are the main sources of catch-up. Second, it can be put that since the catch-up process is shaped by the actual NIS capabilities, an effort to increase this asset and improve its functionality would jump the country's growth trajectory nearer to the international frontier.

The model sketched above accounts in a relatively straightforward manner for the following general property that all efforts addressed to improve the functionality of the NIS capabilities – related to both institutions and resources – by a country could be viewed as a necessary condition to improving its economic performance. In the same sense one can see that a catch-up (or forging-ahead) growth trajectory is taking place. Conversely, a falling-behind trajectory is associated with NIS capabilities that are poorly organized, non-access to international networks and lower efforts to increase these assets.

Conclusion

From international research on innovation and the governance of technology, we know that there is an increasing amount of empirical evidence supporting the phenomenon that the rate of introduction of innovations by firms and sectors seems to be increasingly influenced by the ability and capacity to cooperate with other firms and organizations. Networking activities are, without doubt, key in understanding policy implementations. Therefore it can be drawn from all these new international contexts that one of the determining factors driving the internationalization of R&D is the capacity of firms to efficiently manage decentralized complex systems on a world level. As well, other types of networks – for example, between universities and research centers – need an increasing capacity to manage complex linkages with and between government and international organizations. It is likely that the lack of such a capacity is a main obstacle to the internationalization of research and networks. These observations thus provide a framework for analysis in which the traditional neo-classical and political-economic trajectories for growth and development are deemed largely moot. Questions of technological choice, appropriate technology and transfer, low wage–low skill employment, and so on – all based on dependency theory and hegemony – become de-emphasized in light of innovative practices, knowledge and learning.

The globalization of economic production and knowledge activities might even increase the divergence across nations and regions if local knowledge externalities are not captured by the developing nations. As the number of non-traditional nations becoming active in the performance of science and technology increases, and as regional cooperation pacts and networks are forged by countries across regions in an effort to pool education, labor, innovation and technological resources (as is happening across APEC and the CARICOM), the ability to catch up and forge ahead becomes real. As mentioned earlier, our general conjecture is that national (and regional) innovation systems deeply shape the development pattern of each economy (de la Mothe and Paquet, 1998a; Cimoli, 2000). However, the new opportunity and needs for an increasing accessibility to markets, personnel and knowledge have to be captured too. In this context, the potential conflict derived from the modes of governance of the national system and its international accessibility go together with a specific and new set of incentive structures which should promote the local distribution and diffusion of knowledge. These are the times when the discretionality of modes of governance is the highest and orthodox approaches may produce a development path characterized by a continuous loss of technological and knowledge opportunities.

Notes

1. The work reflected in this chapter was initially carried out by the authors on behalf of CONACYT (Mexico), the APEC Expert Working Group on Regional Cooperation in Science and Technology, and the Organization of American States (OAS).
2. DOA is an acronym for 'dead on arrival' while DEE refers to 'Dynamic Emerging Economics'.
3. Many of these deal with accumulation and can be traced back to Marx's *Das Kapital*, the first volume of which was published in 1867. More specifically, one of the centerpieces of Marx's theory of capitalist development held the view that the rate of profits (and hence of accumulation) inevitably declines as growth takes place. These profits were produced by labor from surplus labor. Competition in the face of this declining rate of profit leads to stronger owners of capital swallowing up the weaker who are then driven into the pool of workers (proletariat). Gerhard Mencsh has also explicated such lines in his *Stalemate in Technology* (1982).
4. See: Metcalfe (1989), Metcalfe and Soete (1984), Amable (1992, 1993), Boggio (1993), Soete and Verspagen (1992), Verspagen (1990, 1991), Dosi and Freeman (1992) and Fagerberg (1995), Dosi and Soete (1983), Dosi, Pavitt and Soete (1990), Cimoli and Soete (1992), Canter and Hanusch (1990).
5. The Abramovitz notion of differentiated 'social capabilities' is quite consistent with this view (Abramovitz, 1989).
6. Other definition of National Technological capabilities which capture the main features of learning and institutional setting is associated to 'complex of skills, experience and conscious effort that enables a country's enterprises to efficiently buy, use, adapt and improve and create technologies. The individual enterprise is the fundamental unit of technological activity, but national capability is larger than the sum of individual firm capability. It comprises the non-market system of linkages, business culture and institutions that enable firms to interact with each other, exchanging the information needed to coordinate their activities and to undertake what effectively amounts to collective learning' (Lall, 1997).
7. There also exists an international dimension given by MNEs' strategies that provokes slipover effects of technology policies of one nation on those of the others. Increasing mobility of firms resources (FDI, strategic alliance, etc.) and the globalizing of the world economy can modify the main elements that explain the origin and the orientation of the NIS-competence, in term whether it is biased in favor of the MNEs' strategy or mainly determined by the country's scientific and technology policy. On the other hand, the national unit is too large to understand the effects of the innovative process in a particular area: it becomes, therefore,

important to focus on the appropriate unit of analysis, and therefore on distinct systems geographically and institutionally localized.

8. In this respect, π is considered as the average labor productivity in the industrial sectors. However, this analysis could be extended to a reconsideration of capital inputs, see Dosi, Pavitt and Soete (1990).

9. In this case that a similar consumption pattern prevails and production costs are given and fixed in both economies.

10. This figure is obtained from a model with a continuum of goods whereas the dynamics increasing returns are introduced, see Cimoli, Dosi and Soete (1986) and Krugman (1987).

11. In this context, the introduction of dynamics increasing returns is assumed to explain the interplay between productivity and income.

12. Most of the works on catch-up put the emphasis on the scope for imitation (Gomulka, 1971, Abramovith, 1979, 1994 and Maddison, 1991). The scope for imitation here is considered as a feature of both competitiveness – organizational and quality improvements – and NIS – capabilities efforts.

CHAPTER ELEVEN

Technology and Employment

Luc Soete

Introduction

The consensus amongst policy-makers, businessmen and even academic economists on the importance of new information and communication technologies for growth and employment appears based on three features: the dramatic decline in the price of information processing, the convergence between communication and computer technology and the rapid growth in international electronic networking. At last, there appears a broad consensus that societies are on the eve of a major breakthrough in the widespread use and application of new information and communication technologies (ICT) throughout all parts of the economy.[1]

The contrast between the above view and the fear amongst the public at large about the benefits of such changes to individual and social welfare is striking. Thus, a large and growing majority of the European public is convinced that 'the computer' will ultimately reduce employment opportunities and threaten job security.[2] Similar concerns about the 'end of work', de-skilling and employment relocation to low wage regions have been raised by a number of popular writers both in the US and Europe (Handy, 1994). Perhaps it is not too surprising within the European context of high unemployment, that many people associate such rapid technological progress and globalization with threats to their social position and less with wider sets of opportunities.

The growing concerns about the employment and distributional implications of the emerging information society are, as has been emphasized in many recent policy reports, not based on any historical precedents. As illustrated by Box 1, taken from the OECD Jobs Study, concerns about so-called 'technological unemployment' have a long track record and one which has not been supported by much evidence. Those predicting the employment implications of a particular set of technologies have tended to overestimate the visible, direct employment displacement effects of the substitution of labor for capital and underestimate the various indirect employment 'compensation' effects which are likely to operate in the economy at large.[3] On the other hand, most macro-economic analyses simply assume the issue away: with proper functioning labor and capital markets, there can only be positive growth and employment effects.[4] The introduction of new technology will increase productivity and hence real wages. If, since the late 1970s and 1980s, productivity growth has slowed down, it is rather the result of a slowing down of technical progress or the lack of new innovations.

Box 1 Technological change and employment: the lessons of history

Forecasts that the next wave of technological change will cause high unemployment and/or large declines in real wages have been frequently made in the past 200 years. Today, these fears take a variety of forms; beliefs that computers and robots will soon take the place of unskilled labor or that the rate of structural change in the economy is too high and will lead to overwhelming dislocations through technological unemployment. So far, all such forecasts have been wrong. Increased productivity has been accompanied by rising demand for labor and rising real wages.

In the 1760s the French physiocrats were arguing that only agriculture was truly productive and that the shift of the French labor force into other sectors would reduce national wealth. As Adam Smith paraphrased their argument, the workings of the market cause the number of service and manufacturing workers to grow over time and 'encroach upon the share [of employment] which ought properly to belong to this productive [agricultural] class. Every such encroachment must ... necessarily occasion a gradual declension in the real wealth and revenue of society.'

Over the following half century the agriculture share of the French labor force continued to fall. But the productivity increases led to average wages in France growing by nearly one-fourth, while unemployment showed no appreciable rise.

In the 1920s the 'Ricardian socialists' were arguing that increased productivity from the introduction of machinery would reduce employment, and put downward pressure on wages and on populations. They were relying on an extended numerical example from Chapter 30 of David Ricardo's revised *Principles of Political Economy*, after which Ricardo concludes 'that the opinion entertained by the labouring class, that the employment of machinery is frequently detrimental to their interests, is not founded on prejudice and error but is conformable to the correct principles of political economy.'

Over the following half-century, capital accumulation and technical change caused average wages in the United Kingdom to more than double, while unemployment showed no appreciable rise.

In the 1860s Karl Marx was writing in *Capital* 'the greater the social wealth, the functioning capital ... the productivity of labour ... the greater is the industrial reserve army [of the employed] ... the more extensive [are] the pauperised sections of the working class.'

Over the following half century, as capital intensity and labor productivity continued to increase, average wages in the United Kingdom once again roughly doubled, and unemployment did not appreciably rise.

In the 1940s, after World War II, cyberneticist Norbert Weiner was forecasting that the invention of the computer would, as Paul Samuelson reports it, create 'technological unemployment that will make the Great Depression look like a picnic'. Over the following 40 years, average hourly wages in the United States more than doubled, while unemployment increased by an average of one or two percentage points.

Source: OECD, 1994

As mentioned already, the technology employment debase is an old one, and while there are some new features to the current debate, it is nevertheless useful to put the current debate in a broader historical perspective. The latter highlights, at least in my view, also how analytically restrictive traditional 'general equilibrium', incremental approaches to technology are with respect to their likely growth and employment impact. Like many others, I would argue that the present-day debate surrounding the emerging information society signals the fact that the widespread diffusion of new information and communication technologies is ushering society into an entirely new era or 'post-industrial' society. The latter, as argued below, should be the starting point for any understanding of the nature of the structural transformation the information society is likely to bring about in our economies. ICTs have a fundamentally different impact on the traditional industrial and industrially dependent service sectors (transport, trade and distribution of material goods) than on 'pure' service activities. This implies, as argued in the third section of this chapter, a continuous shift in value added away from material production and handling toward the immaterial content, as argued in a fourth section. The implications for employment growth and displacement are fundamental. As I argue in the final section, they appear to lay behind many of the growing concerns in Europe about the employment implications of the emerging information society.

Technology, growth and employment: what's new?

The relationship between technology, growth and employment has traditionally been the subject of many contributions in economics.[5] While controversial and the subject of intense debate over the last two centuries, the relationship appears today straightforward at least from the macro-economic perspective. Either the introduction of new technologies leads to more efficient production processes, reducing costs by saving on labor, capital, materials, energy, or any other factor of production, or it leads more directly to the development of new products that generate new demand. In either case, more welfare is created: in the first, through more efficient production combinations that liberate scarce input resources; in the second, by satisfying new wants.

The extent to which this higher welfare or increased productivity feeds back into employment growth depends on the extent to which firms translate productivity gains into lower prices and new investment and consumers respond to lower prices in terms of greater demand. The job losses that often follow the introduction of a now labor-saving process, for example, are compensated by the job creation associated with the output growth following the decline in prices, by additional employment creation in other sectors, particularly the new technology-supplying sector, and by the possible substitution of labor for capital following the downward wage adjustment that clears the labor market.

However, the extent to which new or improved products generate new employment growth depends on whether old products are replaced by new ones and on the responsiveness of consumers to the new or improved goods or services (reflected in the size of the income elasticity of demand). As long as there are unsatisfied needs in the economy and as long as labor and product markets are sufficiently flexible, technological change, even in the form of new labor-saving production processes, does not reduce aggregate employment but generates more growth and jobs.

Most of the controversies that have dominated the economies literature on this issue over the last decades have centered on the automatic nature of the various compensation effects described above. Many contributions have questioned the way in which cost reductions following the introduction of new technologies are effectively translated into lower prices and are likely to lead to more output growth:

the functioning and flexibility of product markets depend in part on the firm's monopoly power, the degree of economies of scale, and various other factors influencing 'price stickiness'. Similar issues can be raised with respect to employment growth and the functioning of labor markets; they range from downward wage flexibility to the many mismatches typical of relatively heterogeneous labor markets. In either case, it is less that technology is at the center of the debate than the speed and clearing function of the product and labor markets.[6] The relevant policy issues fall primarily under the heading of improving the functioning of labor and/or product markets.[7]

Other contributions in the classical economics tradition have questioned the possibility of *ex post* substitutions between labor and other factors of production. At least in the short term, the implications of a more rigid fixed set of production coefficients for analyzing technical change and employment are relatively straightforward. Labor-saving technological change embodied in new investment could, if wages adjust slowly, lead to unemployment because of insufficient investment to maintain the full-employment capital stock;[8] this is the so-called 'capital-shortage' unemployment.[9] There was a lively debate during the 1980s on the extent to which the increase in unemployment in European countries in the 1970s could be due to this phenomenon.

Yet other contributions question the automatic nature of the link between input-saving new technologies and productivity gains. Most of these studies – which often attempt to explain the 'productivity paradox' – are empirical in focus and attempt to find reasonable explanations for the disappointing performance of productivity growth in most OECD countries over the last two decades, despite rapid growth in knowledge investment, in particular in private sector R&D, and the emergence of the new cluster of information and communication technologies.[10] However, the discussion is far from over. In particular, there have recently been a large number of empirical and theoretical contributions from growth economists (e.g., Young, 1995; Mankiw, 1995).

Finally, some recent contributions have focussed explicitly on the international 'open economy' framework within which most compensation mechanisms are likely to operate. As a result, the relatively straightforward linkages between technology, productivity growth, and job creation mentioned above appear much more complex. A relatively simple elaboration in terms of employment compensation due to foreign demand, e.g. through export and import elasticities, complicates the matter greatly (Stoneman, 1984). More complete pictures including not only trade but also the effects of international spillovers of technology on productivity growth or international capital mobility make it much more difficult to identify the key links between the introduction of a new technology and the ensuing domestic employment impact.

Many of the recent concerns about the implications of technological change for employment appear to relate to these international compensation mechanisms and to the way that gains from technological change are distributed internationally. In the gloomy vision of some popular authors,[11] 'wages in the most advanced economies are being eroded owing to the emergence of a global market-place where low-paid workers compete for the few jobs created by footloose global corporations' (Rifkin, 1995). In other words, the globalization of industry and, increasingly, of services casts a new light on the interaction between technology and employment in an open economic framework characterized by low transport and communication costs.

While it is still generally agreed that in a 'world' economic framework, input-saving technical change leads, through increases in productivity, to higher welfare, wages, and growth and thus generates new employment, the impact on individual countries is now much more complex and is based on a broad range of macro-economic and micro-economic adjustment mechanisms. This is especially because positive and negative effects do not coincide either in time or in space: adjustment takes time, and the industries and types of workers that will benefit from technical change are different from the ones that lose from it.

At the same time, the premium placed on the role of knowledge and on the acquisition of skills in this global environment implies that international differences in the pattern of employment and unemployment in industrialized economies may be coming to depend increasingly on the capacity of national economies to innovate, enter new, unregulated 'service' areas and/or absorb new technology more rapidly.

The information society: a move toward a post-industrial society?

Contrary to common belief, the notion an 'information society' is not new. Its origins go back at least to the earlier concept of a 'post-industrial society'[12] and the subsequent recognition of the importance of 'information technology' in speeding up this process of industrial transformation.[13] The rediscovery of the concept of 'information society' today, is in the first instance inspired by the continuous speed of technological improvement (historically unprecedented) and the extraordinary pervasiveness of information and communication technologies.[14]

In doing so, the policy debate has, not surprisingly, been dominated by the technological and infrastructural requirements, including the necessary changes in the regulatory framework. Both issues have been extensively researched in all of the G7 countries. In Europe there is now broad and general policy convergence on the essential need for common technical standards, open access and rapid liberalization of telecommunications. However, there remain fundamental questions about the speed of regulatory reform which typically appears out of scope with the speed of the technological improvements in the ICT area. Furthermore, the debate has remained dominated by supply side issues.

As the High Level Expert Group on the Social and Societal Aspects of the Information Society[15] stressed, it is essential to enlarge this debate to the demand side. It suggests a return to some of the issues raised in the older information society debate: the shift, implicit in the notion of 'information society', to a 'post-industrial' society, in which the largest part of new value added is being created in the non-industrial sectors of the economy. The growth of the tertiary (service) sector, and within that, of what can be called the information services sector, is one of the most important economic and social phenomena of Europe's current industrial transformation. In fact, it is likely to be one of the most important structural change features of the next century.

Before discussing these, it is important though to 're-situate' the current information and communication technology transformation and how it has affected manufacturing and services.

New information and communication technologies: bridging time and distance

The dramatically increased capacity to store, process and disseminate information at minimal cost has been described most extensively in the context of industrial (or agricultural) production processes. Pre-dating even the early 'information technology' literature, the so-called 'automation debate', popular in the US in the mid-1960s, described how labor-saving 'robotics' would raise industrial productivity and bring about major organizational changes. In line with this literature, many IT analyses have always wondered how confronted with such pervasive cost-reducing technologies, economies would be able to generate sufficient new employment (the various price and substitution elasticities being too low to bring about sufficient employment compensation).[16] More recently, the specific impact of new information technology on services has re-entered this debate. It could be argued that the impact on services will be more of an opposite nature compared to the impact on manufacturing.

In many ways services can be defined[17] as those activities (sectors) where output is essentially consumed when produced. While this might well be considered a rather narrow definition and one which covers only a limited number of sectors presently falling under the statistical definition of service sectors, it is an analytically useful definition because it highlights the intrinsic immaterial, intangible nature of many service activities whether they are personal services, such as hair-cutting; entertainment such as an opera performance; education such as teaching; health such as a doctor's visit; or public services such as applying for welfare services. It is this similarity feature of production and consumption which has generally limited productivity improvements in such activities. In parentheses, it is also worth noting that this also provides an intellectual argument as to why the economic profession has generally tended to ignore the study and analysis of service activities.

Information and communication technologies, almost by definition, allow for the increased tradability of service activities, particularly those which have been most constrained by the geographical or time proximity of production and consumption. By bringing in a space or time/storage dimension, information technology will make possible the separation of production from consumption in a large number of such activities, hence increasing the possible trade of such activities.[18]

In the case of the current 'new' information and communication technologies and their potential not just to collect, store, process and diffuse enormous quantities of information at minimal costs but also to network, interact and communicate across the world – the world as a global village – both the time/storage and space dimensions of the new technology are likely to bring about the further opening-up of many service activities, increasing their domestic *and* international tradability. As in the case of the telephone, it is likely that the 'new' emerging computing and other electronics manufacturing sector will in the end remain relatively small compared to the growth and size of new information and communication service sectors. However, even the definition of the latter will become blurred as more and more of the traditional 'physically present' service activities will become 'info-type' service activities.

With regard to more traditional production processes, typical of industrial production, but also common in traditional service sectors such as transport, wholesale and retail trade, the impact of ICTs could well be characterized as exactly of the opposite kind. Rather than bringing time/storage between production and consumption as in services, ICTs will in the first instance aim at *reducing* the time/storage dimension between production and consumption. Many of the most distinctive characteristics of the new information and communication technologies are related directly to the potential of the new technology to link-up networks of component and material suppliers, thus allowing for reductions in storage and production time costs – typified in the so-called Just-in-Time production system. At the same time, the increased flexibility associated with the new technology allows for a closer integration of production with demand, thus reducing the firm's own storage and inventory costs – which could be typified as Just-in-Time selling. Both features clearly work in the opposite direction from what was said above, i.e. they aim at reducing the *time/storage* dimension between production and consumption. In doing so, they might well reduce the 'tradability' of a number of those intermediary storage and inventory activities.

In essence therefore, ICTs are making services more tradable and more like manufacturing, leading to a further convergence of industrial and service activities.

Information service activities: the new economic driving factor?

With the growing convergence between manufacturing and services, service activities (corresponding on average to two-thirds of economic activity in the EU) appear increasingly as important in their own right and in an increasing number of areas, and to be dominating manufacturing rather than the other way round. Particularly since the emergence of ICTs and their 'tradability' impact on many service activities, the latter activities have emerged as 'core' value-added activities. It is essential in this context to distinguish amongst data and information and knowledge.

ICTs play an essential role in the 'codification' of knowledge. Codification implies that knowledge is transformed into 'information' which can either be embodied in new material goods (machines, new consumer goods) or be easily transmitted through information infrastructures. It is a process of reduction and conversion which renders the embodiment or transmission, verification, storage and reproduction of knowledge especially easy.[19] In contrast with codified knowledge, tacit knowledge refers to knowledge which cannot be easily transferred because it has not been stated in an explicit form. One important kind of tacit knowledge is skills. The skilled person follows rules which are not known as such by the person following them. They are linked to activities acquired through learning but often of a non-routine kind.[20] The most important impact of new ICTs is that they move the border between tacit and codified knowledge. They make it technically possible and economically attractive to codify kinds of knowledge which so far have remained in a tacit form.

The embodiment of codified knowledge in material goods has been typical of the dramatically increased performance of many new capital and consumer goods, incorporating many new electronic information and communication devices. The latter in turn have been at the core of the continuous productivity, investment and consumer demand growth in Western societies. As emphasized by authors criticizing the early 'post-industrial society' literature,[21] this process could also be described by a process of 'industrialization' of services: the continuous replacement of particular service activities by household material goods, embodying at least the 'codified' knowledge part (washing machines, televisions, dryers, etc.). The more recent electronic improvement in these products has further increased the 'household' performance of these products, freeing further household time. While the quality of these new material goods will not always substitute for the service activity they replace (a dishwasher is a good example), the codification process will be to some extent complete. The product might lack user friendliness (the typical example being the video player), but the user is not required to possess, or to understand, the knowledge embodied in the machine.

In services, by contrast and following the arguments set out above, while the codification of knowledge will have made such knowledge more accessible than before to all sectors and agents in the economy linked to information networks or with the knowledge how to access such networks, its immaterial nature will imply that the codification will never be complete. The codification process will even rarely reduce the relative importance of tacit knowledge in the form of skills, competencies and other elements of tacit knowledge, rather the contrary. It is these latter activities which will become the main value of the service activity: the 'content'. While part of the latter might be based on pure tacitness, such as talent or creativity, the largest part will be greatly dependent on continuous new knowledge accumulation – learning, which will typically be based on the spiral movement whereby tacit knowledge is transformed into codified knowledge, followed by a movement back where new kinds of tacit knowledge are developed in close interaction with the new piece of codified knowledge. Such a spiral movement is at the very core of individual as well as organizational learning.

The implications for this continuous shift in value from manufactured goods embodying increasing amounts of 'codifiable' knowledge towards service based 'tacit' knowledge activities is typical of the new emerging information society. It explains the attempts of electronic and computing manufacturing firms to enter information content activities. Within services, it explains the move of 'carrier' operating firms being most directly confronted with the codification of knowledge and its distribution, to enter content sectors (media, education, culture).

The emerging information society will in other words become crucially dependent on user demand for new information products and services. The 'demand articulation' of the latter depends crucially on the existing regulatory institutional framework as well as the overall macro-economic climate. Much has been written about the former. The McKinsey Global Institute study points to existing (over)regulation in Europe in a number of typical 'information society' service sectors, such as multi-media, preventing these markets to emerge more rapidly. The emergence of new service markets involves, practically by definition, new regulatory frameworks. In many typical cases of information or communication services, their commercial success critically depends on the reaping of the substantial economies of scale involved in such typical network activities, with marginal costs being a fraction of fixed costs (one may think of movies, software, financial and insurance services, etc.). Europe with its fragmented national markets and cultural diversity is clearly at a disadvantage in reaping such economies of scale. The various proposals for liberalizing and harmonizing some of these markets more quickly, and in particular the national telecommunications operators' markets, as for example in the Bangemann Action Plan or the G7 Industrialists' Action Plan, correspond to the urgency felt in addressing at least a part of this problem.

The new employment challenges

The particular employment concerns associated with the emerging information society, relate in the first instance to the likely impact of ICT on productivity growth and output growth, particularly with respect to some of the new information service sectors described above. While much will depend in this latter case on the new needs and markets the new information service sectors will be capable of addressing, much will also depend on the way the productivity gains will be redistributed throughout the economy. These latter gains are as much based on the overall counter-inflationary effects of falling costs and prices in micro-electronics, computers and telecommunications which are affecting a widening range of products and services, as on organizational improvements and other more dynamic learning efficiency improvements at the level of the shop floor, production planning or administrative activities.

In attempting to assess the employment creation and destruction effects of ICT it is therefore no longer possible to distinguish many of the direct negative and positive effects from the indirect effects. The direct effects represent as much the new jobs in producing and delivering new products and services, as the old jobs being replaced by new ICT equipment. The indirect effects are the amalgamation of the many positive and negative consequences elsewhere. Thus while computer terminals are everywhere, it is not always clear whether they are displacing workers or adding additional services and employment. The computer industry itself provided machines which displaced earlier types of electro-mechanical office equipment, whilst the micro-electronic industry largely displaced the old valve (tube) industry. The new digital telephone exchanges require far less labor to manufacture and to maintain than the old electro-mechanical exchanges and the number of people working in the telephone switch industry has fallen in most industrial countries. Competitive restructuring of the old monopolistic networks has also resulted in a reduction of the number of employers, even though the number of firms and the number of lines and cells has increased. However, the new telecommunications infrastructure provides the basis for numerous new information service industries and equipment, such as e-mail, facsimile, data banks and the new multi-media services of the future.

To compare the balance of gains and losses is therefore a difficult undertaking as numerous empirical studies of the 1970s and 1980s confirm. The naive view of ICT as simply a process of automation and job destruction has its counterpart in the equally naive view of ICT as a purely positive source of new employment.

What brings employment concerns about the information society today back to the public policy forefront, despite the reassuring historical arguments and macro-economic compensation arguments, are the particular features and characteristics of the new information and communication technologies underlying the IS, reviewed above. I would emphasize five of them.

First, there is, as implicit in the arguments set out above with respect to ICTs and services, the particular impact of new ICTs on employment in the service sectors, particularly in those sectors and occupations hitherto largely 'protected' from automation or 'informationization'. In so far as such 'sheltered' service employment has acted in the past as the main absorber of employment displacement in manufacturing and agriculture, there is increasing concern about whether new services will indeed be capable of providing sufficient new employment opportunities. Such new services are crucially dependent, as highlighted above, on the appropriate regulatory framework. The emergence of new markets for information services requires indeed not just a more competitive framework, deregulation and open access, but new institutions setting out so to say the rules of such new markets, including those governing property rights, security, privacy, etc. At the same time, and as the case of the Internet illustrates, the speed of change goes beyond the 'controlled' liberalization process pursued at the moment in Europe, and involves a much more dramatic 'creative destruction' process, with a completely new communication pricing structure. Particularly in Europe, there is, in my view, justified concern that the regulatory reform is too slow and that the development of new services in Europe is consequently lagging behind the development in other parts of the world.

Second, in so far as the dramatic decrease in the cost for obtaining data and information with new ICTs can be compared with a macro-economic deflationary effect, the question can be raised whether our present statistical methods for assessing 'inflation' are still appropriate, and not increasingly measuring inflation 'illusion' rather than money illusion. As I have argued elsewhere,[22] it seems reasonable to assume that current European estimates of inflation ignoring many new products and services and ignoring many of the quality improvements typically associated with new ICTs, are overestimating inflation by two to three percent a year. As a result, and as one would expect on the basis of the widespread diffusion of ICTs and the emerging information society, many European economies are no longer in a situation of inflation, but rather deflation. The current strict monetary policies implemented in most member countries within the framework of the EMU convergence criteria, based on overestimated inflation estimates, are consequently unlikely to stimulate in the short term the positive output and employment growth aspects of the emerging IS. This is not to deny the possible *longer*-term positive impact of budgetary fiscal consolidation policies on interest rates, private investment and employment growth – so-called crowding in – as argued, for example, in the Commission's European Employment Strategy. However, in the short term one may indeed wonder whether the priority given in the EU to strict macro-economic monetary policy is likely to be conducive to a rapid emergence of the information society in Europe.

Third, since ICT is an 'information' technology – the essence of which consists of increased memorization and storage, speed, manipulation and interpretation of data and information – it has and will further increase the possibilities to 'codify' large parts of human skills. As emphasized in the previous section, this is not to deny the importance of the 'tacit' part of knowledge. On the contrary as more knowledge becomes codifiable, the remaining non-codifiable part is likely to become even more crucial. Thus, the ability to codify relevant knowledge in creative ways as well as the competence to sort out relevant information and to use it efficiently has and will continue to become of much greater importance. By the same token though, an increasing number of routine skills are becoming totally codifiable and their importance dramatically reduced.[23] As the largest part of employment in our societies does involve such routine tasks, there is increasing concern about the distributional employment impact of the information society. Furthermore, confronted with the accompanying widespread use of various forms of information and computer technologies, 'skill mismatches' are likely to be of a much more pervasive and general nature, raising questions about the inherent 'skill bias' of new information and communication technologies. While these distributional concerns point to the crucial need for the broadening of education and training needs to all groups in society, they also raise fundamental questions with respect to possibly excluded groups, such as unskilled or routine skilled labor.

Fourth, as a consequence of the increased potential for further international codification and transferability, ICT could to some extent be considered as the first truly 'global' technology. The possibility of ICT to codify information and knowledge over both distance and time, not only brings about more global access, it also enables firms/organizations to relocate the sort of routine activities which can be codified and thus also internationally traded. In other words, ICT contributes to economic transparency and, in so far as it brings to the forefront the cost advantages of alternative locations, to international capital mobility and international 'outsourcing' of particular activities. While the benefits to the world as a whole of such a more transparent, borderless world, are undisputed, there is, again in my view, justified concern about the worldwide distribution of those benefits. For the poorest, most peripheral countries/regions there is concern of becoming excluded; for the richer, technologically leading countries/regions, there is concern about the increasing erosion of the monopoly rents associated with innovation, and their implications for employment, wages and the built-up social security systems in Europe. In most EU countries, the financing of the national social security system and more generally the welfare state has been closely linked to employment, both through contributions from the employer and employee. In an increasingly global and economically transparent world, such a national link is becoming undermined, raising fundamental questions about international competitiveness and/or the long-term sustainability of the national welfare system.

Fifth, much has been written about the likely growth and employment impacts of EMU, in particular the increased transparency in costs and prices, increased financial mobility thanks to the harmonized currency, better guarantees with respect to European-wide price stability and last but not least the crowding in effect of private capital thanks to the fiscal consolidation resulting from aiming for the public debt Maastricht criteria. Little has, however, been written, let alone researched, on the likely impact of a set of radical new technologies such as ICTs on employment growth in a monetary unionized EU. A rapid and frictionless introduction of new ICTs, of accompanying new organization methods as well a better and a more EU integrated institutional framework will undoubtedly all be essential factors for European competitiveness as well as the internal competition between European regions. Alongside better 'natural' locational advantages and infrastructural provisions, the use of more efficient production techniques and organization forms, including local networking with suppliers and users, will become the essential factor behind the maintenance of a high income local economy. At the same time, maintaining such 'local competitiveness' will have to be maintained to a larger extent than before through the building up of international, extra-EU competitive strengths: despite the availability of a large, now also monetary harmonized market, presence at the world global level will be essential for European competitiveness. How the currently renewed importance given to technology and more generally knowledge acquisition and ICT diffusion both for internal and external competitiveness can be achieved within a European monetary union, characterized by nationally, often even regionally segmented labor markets, will become the core issue for European policy-making in this new century.

Conclusion

It is difficult to predict the precise employment impact, whether in terms of volume, sectoral or occupational composition, of the new information and communication technologies and the emerging information society. Both direct and indirect employment effects are likely to be substantial and will undoubtedly further intensify the structural transformation of society. Thus as the rise of entirely new industries in the second half of this century, such as the software industry, the electronic computer industry itself, the micro-electronic industry, the VCR and the TV industry, has brought about new employment opportunities – with each of these industries which barely existed before 1950, now employing millions of people – so will new information. and communication service sectors emerge and provide new employment opportunities. However, to do so will depend crucially on whether the appropriate regulatory institutional environment exists for the emergence of new information and communication markets; and on the appropriate macro-economic climate. On both accounts there it concern that present day conditions in the EU countries are unlikely to favor the rapid emergence of such new markets.

At the same time, there is again, particularly in Europe, apprehension that both the direct and indirect employment displacement effects of the information society will be substantial. The potential offered by the new ICT to increase efficiency in some of the most typical information and communication dependent services sectors, such as finance, insurance and other business services, is high: the number of jobs displaced particularly of the routine kind is expected to be high in those sectors. Furthermore, there is concern that with the advent of the information society many traditional, 'non-tradable' service activities might become internationally tradable and that those activities will be relocated in low wage countries or regions. Again those jobs likely to be displaced are the sort of simple routine jobs, not necessarily of the unskilled, manual type, but rather those in which the information or knowledge content appears most easily subject to codification through ICT.

Whether the net balance of these direct and indirect employment effects will in the end be positive or negative cannot be assessed by simply counting new jobs gained and old jobs destroyed. It has, in my view, to be recognized that the expansionary effects on any national economy or the world economy as a whole of the information society will paradoxically depend on the way the new information and communication technology will create the basis for a virtuous circle of growth in which investment is high, labor productivity grows fast but output grows even faster, so that there remains a net growth in employment. That is why I remain fundamentally optimistic about the inherent, long-term welfare and new employment opportunities associated with the emerging information society. However, as also emphasized in the HLEG report these welfare gains and new opportunities are not given. Whether the virtuous circle described above can be sustained in Europe will depend on macro-economic policies, regulatory and institutional reform in labor and product markets, more user friendly and focussed technology policies, new distributional policies, policies aimed at boosting productivity growth as well as policies aimed at 'integrating' the new information and communication technologies in society. If there is a good match between technologies, policies and institutions, prolonged periods of full employment are much more likely to result.

Within the framework of the EMU these developments raise indeed new policy challenges directly related to various aspects of the economic use and organizational integration of new technologies, all of which are likely to have direct or indirect implications for employment creation at the EU and local level, and all of which are likely to become crucially dependent on successful 'skill matching' at European and local level. Two such challenges are likely to dominate the European policy debate in the years to come.

First and foremost, there is likely to be increased friction about the demand and supply for various categories of skilled labor between regions and countries. There are likely to be increasing growth bottlenecks associated with the unavailability of particular skills, competencies or knowledge in particular regions and not in others. The lack of mobility across Europe is likely to be felt much more directly in a monetary harmonized Europe than in a free trade, single market Europe.

Second, there is likely to be also increased productive capital mobility across Europe. Such mobility will increasingly become dependent on 'real' factors such as the ratio productivity/gross wage cost (unit labor costs) or the local availability of other crucial production factor inputs (access to qualified personnel, qualified suppliers, infrastructure, etc.). In both cases the costs and availability of skilled labor will become crucial 'locational' advantage variables.

Notes

1. As Al Gore put it at the opening of the Brussels G7 summit: 'Just as human beings once dreamed of steam ships, railroads and superhighways, we now dream of the global information infrastructure that can lead to a global information society.'
2. See Eurobarometer survey (1992).
3. These include output compensation effects depending on price and income elasticities, indirect employment creation effects in the new machine, capital goods producing sector, as well as possible substitution effects following wage adjustments.
4. This view is the one most clearly reflected in the OECD Jobs Study: it is the over-regulation of European labor markets which is at the core of the present European unemployment problem. By contrast, in the more recent submission to the Lille Jobs summit, the OECD report on Technology, Productivity and Employment Creation puts forward a more sophisticated analysis, enlarging the policy debate beyond the usual call for more flexible labor and product markets. See OECD (1996).
5. There is now a voluminous empirical and theoretical literature on this subject. In this

century alone, we would distinguish four sets of economic debate on the relationship between technology and employment. The first, probably the most classical in its origins, took place during the economic depression of the 1930s. Contributors included Hansen (1931, 1932), Kaldor (1933), Weintraub (1937), and Neisser (1942). Many of the issues and concerns raised by these authors sound quite familiar today, particularly in the context of the notion of increasing returns in current 'new' growth models (e.g. Aghion and Howitt, 1991). The second debate focussed mainly on the postwar United States and the fear of 'automation'. In the 1960s, levels of unemployment were higher in the United States than in Europe, and many blamed technological change. As a result, a National Commission on Automation was appointed and produced a massive six-volume report (US National Commission, 1966). This debate had little influence in Japan and in the European countries that were rapidly catching up. The third debate, which began in the late 1970s, was particularly active in Europe. It focussed on the emergence of the cluster of computer-based communication, information, and automation techniques associated with microelectronics, which appeared at first glance to have great labor-displacing implications (e.g. Freeman *et al.*, 1982; OECD/ICCP, 1982; Katsoulacos, 1984). The fear that those displacement effects might dominate the compensating job creation effects for quite some time recalled in many ways the classical debate. As there, it appeared to be a reflection of the times: there was a set of 'revolutionary' new technologies and persisting high unemployment. The most recent upsurge focusses much more on the global aspects of the new information and communication technologies and the possible erosion of employment and high living standards in the advanced countries. Originating to some extent in the United States, and linked to the political debate surrounding NAFTA, it quickly 'globalized', and contributions have come from throughout the world.

6. As von Mises (1936: 485) put it: 'Lack of wages would be a better term than lack of employment, for what the unemployed person misses is not work but the remuneration of work. The point [is] not that the "unemployed" [cannot] find work, but that they [are] not willing to work at the wages they [can] get in the labor market for the particular work they [are] able and willing to perform.'

7. Both the OECD Jobs Study (1994b, 1994c) and McKinsey Global Institute (1994) can be said to have focussed primarily on these market issues. The former emphasized the functioning of labor markets, the latter the functioning of product markets, particularly in services.

8. This result is obtained by Venables (1985), for instance, with the use of general equilibrium setting with fixed coefficients.

9. '"Capital shortage" unemployment reflects a seeming lark of "productive" capital to employ part of an "adequately" skilled and suitably located labour force. ... Capital shortage unemployment can ... occur as a result of both lack of physical capacity and economic obsolescence; if variable costs exceed price, capital will not be used (and the corresponding jobs will disappear) even though such capacity could be operated in a physical sense' (OECD, 1983).

10. In previous reports for IBM, written with Chris Freeman, this issue was already addressed in quite some detail, see amongst others Freeman and Soete, 1985 and 1993. See also the various contributions for the OECD (the so-called 'Sundqvist Report', 1986; the subsequent 'Technology and the Economy Programme', 1992 and the most recent G7 Technology, Productivity and Job Creation project, 1996).

11. See, for example, Aranowitz and DiFazio (1994) and Rifkin (1995). In many ways, and as noted by some trade economists (Krugman and Venables, 1994), such views are to some extent reminiscent of the old Prebisch-Singer dependencia arguments, but applied to the advanced countries. In the old core-periphery models, 'immiserising' growth in the developing countries would take place because all the benefits of increased efficiency gains in raw materials, agricultural, and labor-intensive manufacturing production were passed on to the advanced economies, e.g. through lower prices or higher repatriated profits. In the current view, the

pattern is the opposite: most of the benefits of technological change are passed on to some of the rapidly industrializing countries through more rapid international diffusion of technology from the advanced countries, the reinvestment of profits and relocation of production to those industrializing countries, and the erosion of various monopoly rents in the advanced countries, including wages. In principle though, and in contrast to the Prebisch-Singer model, such a redistribution process should lead, as trade theory would predict, to the convergence of growth and income.

12. See e.g. Daniel Bell's *The Coming of Post Industrial Society* published in 1973.

13. See amongst others Porat's *The Information Economy* published in 1977, Nors and Minc's *The Computerisation of Society* published in 1978 and Tofler's *The Third Wave* published in 1980.

14. As I have argued with Chris Freeman in a number of papers and reports for IBM since the early 1980s. See amongst other Freeman and Soete (1985, 1993).

15. See the High Level Experts Group's Interim Report: Building the European Information Society for us all, Maastricht, 1996.

16. Interestingly, this is also the main argument of those studies in this area which limit their focus to manufacturing (Pianta, 1995; Reati, 1995).

17. For early analyses along these lines see Quinn (1986) and Soete (1987).

18. This was certainly the case with regard to the invention of printing in the Middle Ages and the impact this first new information technology had on the limited tradable 'service' activity of monks copying manuscripts by hand. It was the *time/storage* dimension of the new printing technology which opened-up access to information in the most dramatic and pervasive way and led, to use Marx's words, to the 'renaissance of science', the growth of universities, education, libraries, the spreading of culture, etc. This opening-up, 'tradability' effect would become of far more importance to the future growth and development of Western society than the emergence of a new, in this case purely manufactured-based, printing industry.

19. See in particular David and Foray (1995).

20. One might think of such activities as gardening, biking or housekeeping.

21. See amongst others Gershuny (1978) and Gershuny and Miles (1982).

22. See Soete (1996a).

23. It is interesting to observe that by contrast, relatively simple human tasks (such as gardening) might never become codifiable. This explains why the idea of ICT as a 'skill biased' technical change, does not really capture the complexities of the de- and reskilling processes. See in more detail OECD (1996).

CHAPTER TWELVE

Governance Challenges of Technological Revolutions

Francis Fukuyama and Caroline Wagner

Introduction

Emerging technologies are presenting unprecedented governance challenges to the international political system. These technologies are: (1) electronic communications and computer intelligence emerging from the information revolution, (2) human genetic manipulation and bio-informatics emerging from the biotechnology revolution, and (3) nano-technology emerging from the quantum revolution. How policy-makers respond to the challenges raised by these technologies, including the extent to which developments are supported with public research funds and whether they are regulated, are matters of increasing concern among citizens and for governing bodies. Existing political institutions are being pressed to deal with difficult questions and challenges raised by these technologies. New governance mechanisms, particularly on an international level, are needed.

The promise and challenge of emerging technologies

The potential shortfalls of the current political and legal infrastructure to deal with global changes in information, biotechnology, and nano-technology are already evident as the Internet evolves into a global information network and develops beyond original information sharing into a global commercial trading system. Electronic commerce is straining existing trade regimes, protocols for the protection of intellectual property, and concepts of currency. It is creating problems and jurisdictional issues for taxation and regulation. A number of governance issues have already arisen, such as the fairness of the current system for allocating domain names on the Internet on an international basis.

In the future, there will be other information security-related problems that will require a high degree of international cooperation to govern or resolve. These include the uses of the Internet for crime, and the misuses of the network by public and private groups in ways that invade personal privacy. Some suggest that contractual relationships will replace regulation and trade protocol. What organization is capable of negotiating and implementing new rules or enforcing net-based contracts? What court of law will adjudicate international contracts agreed to on the GII? Can these simply be folded into the World Trade Organization (WTO), or do they need a separate institution?

Recent developments in biological sciences, particularly genetics, raise the question of international organizational and legal governance. Procedures that are judged ethically or medically objectionable in one country may become available, through market mechanisms, in another location, leading to the development of offshore sites where individuals may go to avoid regulations. Can existing organizational structure and law be adapted to the products of biological science? Will new forms of political organization and law be needed to address these changes?

Similar questions will be raised as biological sciences and computer sciences converge into applications called 'bio-informatics'. As science explores creating information technology that can be used as a human prosthetic – either worn on the body or implanted under the skin – questions about when it is appropriate to use these technologies and under what conditions will arise. Science is also exploring the use of biological materials as information processors in objects such as 'biochips'. Technologists suggest that miniature biological sensors detecting chemical and biological information may soon be available, capable of providing instant feedback on individual or group activities, and further, linking this information into ultrascale networked computing. How can abuses of these technologies, such as surveillance and large-scale information gathering among the population, be anticipated and regulated or countered?

Questions about who within society has the competence to regulate products and processes emerging from new technology will also apply as nano-science develops technologies that can manufacture products from the atomic level. Whether this science has the potential to actually change the nature of production from factory to laboratory production remains to be seen, but the implications of such a change are as great as those posed by information and biotechnology.

What is technology

Technology – the tools that we use to extend our human capacities – has been intertwined with and used to define human history. The technology dominating a particular historical age is often used as a way to define and bound that age – the Stone Age, Iron Age, the Industrial Age, the Information Age. Technologies are designed to alter the natural world and make it more habitable, safer, more productive, or more pleasant. As such, it is an extension of human intention to change the environment in some way. Technology can be developed by individual action, such as carving an arrowhead, or by collective action, such as building a dam. Technology may require little in the way of social or political support, or like the train system, it may require a complex socially developed infrastructure in order to operate.

As technology has evolved, it has moved from simple tools to more complex tools. Complexity is increased as simple tools are combined together into simple machines such as pulleys or gears. The ability to use simple machines often requires learning and skill combined into technique; such learning is passed from one individual to another and becomes tacit knowledge in the heads of users. Simple machines increase productivity and capacity by orders of magnitude more than simple tools do. The capabilities of production, and the resulting products, created the conditions in the Middle Ages for social organization into towns and cities. The know-how to use simple machines was often passed along within a guild or professional society, which, in some cases, also oversaw quality.

With a combination of advances in engineering and science, simple machines evolved over time into more complex machines. The first of these types of complex machines might be called 'active' machines – they include multiple parts in which some parts move and some remain stationary. Clocks, steam-driven boilers, tractors, and so on, combine simple tools and simple machines, often adding autonomous power to create machines of precision and efficiency. The use of active machines required individual skill, but, eventually, also required collective, social, or technical knowledge that extended beyond tacit, skill-based knowledge that could be passed from individual to individual in an informal way. Active machines introduced the need for standardization of practice and the teaching of skills and theory. Boilers, for example – which have a tendency to explode under certain conditions of use – saw the first organized efforts by users to standardize practice. (Cochrane, 1976.)

As knowledge about tool building and use combined into more complex social knowledge and became embedded in knowledge systems (e.g., collected in books and manuals) the capacity to develop more complex machines emerged, as well. Reactive machines were developed in the nineteenth century – machines that had more than one function and whose complex function could be changed mid-course in response to human intervention. Reactive machines include automobiles, machine tools, telephony, etc. These machines not only required a greater degree of infrastructure and standardization of design and use than simple and active machines, they also were the first types of machines to see the introduction of social or governmental regulation over their development and use. (Cochrane, 1976.)

Interactive machines and other types of technology, combined into systems, now dominate the technology landscape. Interactive machines require even more collective skill and management, and conversely, tend to require less individual knowledge to use than active or reactive machines. Interactive machines appear to be 'smart': they reduce the need of the user to know in-depth technical information in order to use them.

At each juncture of development and evolution of technology, there is a period of social and personal discomfort with the new product. Some reactions, such as those demonstrated by the nineteenth-century Luddites, have been violent.[1] While some people can adjust easily to the new, for most people and groups of people, assimilating new technology takes time. It does not always happen that new technologies are accepted, and new tools or new ways of using tools are not always adopted. Generally, technology must extend a human capacity and enable human intention in order to 'fit' the culture and be adopted.

After a period of adjustment to and assimilation of new technology, and once individuals learn new skills or adopt new technology, these items recede into the background of consciousness. A simple example is the skill of riding a bicycle or using electric lights. The same phenomenon of assimilation and acceptance is true of societies as well as for individuals. People in the nineteenth century marveled at the speed of trains. Today, even the fastest trains do not receive much mention.

The term 'technology', when examined critically, is often used to refer to those tools with which society or individuals are not yet comfortable. The more complex a machine is, and the more that collective knowledge is needed to sustain it, the more time it takes to assimilate. The more infrastructure required to sustain a technology, the more likely that governance structures, whether public or private, are created to oversee or support the technology.

What does this mean for the technologies of the twenty-first century? Information, biological, and nano-technologies are complex and interactive. In the case of information technology, the extent of social and technical knowledge needed to sustain the information system is large, as is the infrastructure required but manipulation of the 'product' is in the hands of the individual. Government regulation and private sector standardization have been highly active in telecommunications (allocation of the spectrum, for example) but less so in information technologies. The rate of adoption has been relatively quick, perhaps because the skill of communicating does not need to be learned anew in order to use the new tools.

Based on the definition of technology discussed here, one could question whether biotechnology is a 'technology.' It has anomalous features that make it hard to categorize. It is certainly interactive, but it is not necessarily complex. Recombinant DNA, for example, while resulting from a complex technique, is a simple product. Moreover, it is not a tool, it is a biological agent – it acts on its own once it is 'programmed' to perform in a certain way. The development of biotechnology products requires extensive social and technical know-how, but it does not necessarily require a large infrastructure in order to deploy it. It is not clear what government regulation is required to support or control it (or even whether it could be controlled), and it appears that private sector standardization efforts have yet to emerge in any real way. The practice of biotechnology falls more nearly into the definition of technique – the skill for doing something – rather than a tool itself. Tools are used to make biotech products, and the outcomes are certainly intended to extend human capacities and make the world more habitable, but the nature of the product itself and the way it works sets it apart from other kinds of technologies.

Information and biological technologies and eventually tools resulting from the nano-sciences have in common that the individual is more in control of the use and application of these technologies than is true of many active and reactive machines, where the systemic nature of the technology often requires collective action to put into use. The fact that collective action is not required to use these technologies makes them particularly difficult to govern. The Internet is hailed by many for the 'free-wheeling' way in which it operates – created from the bottom-up and with little governance or control. Indeed, the Internet is often cited as a promoter of 'true democracy' because it enables the individual to interact with others without a socially constructed medium. Biotechnology, too, is seen as having special promise because it will tailor treatments and medicines to the individual and place control of certain biological processes in the hands of individuals.

The features that make these technologies different – and that make their effects orders of magnitude greater than those of other technologies that have emerged in the past 50 years (with the possible exception of that of nuclear weapons) – also make the effects of their abuse potentially greater than those of other technologies. Yet, the level of control that is in the hands of the individual makes social governance much more complex than for technologies that require collective action to build, use, or maintain. The problem that emerges is no longer to ensure democratic control over a large and complex centralized system but rather to determine how much governance is necessary for a decentralized, distributed system and how and whether society can govern it.

The following section describes the key features that make these technologies both a promise and a threat.

Electronic communications and computer intelligence

Electronic communications tools have developed over the past 50 years to the point where they are widely assimilated into society. Even so, social organization is still adapting to the use of these technologies. Like other technologies before it, social adjustment and acceptance of electronics communications is lagging technological development.[2] Electronics technology will continue to grow in speed, ubiquity, and the ease of use. The cost of assessing communications will continue to drop.

The growth of the use of networked communications has been a phenomenon in itself:

- In 1994, there were about 1000 web pages in existence in the world: in 1999, there were more than 400 million web pages;
- In 1993 there were five million Internet users: by 1998 the number was nearly 100 million;
- In 1992 there were ten World Wide Web servers: in 1999, there were perhaps over five million;
- In 1999, traffic on the Internet is said to have doubled every 100 days. Projections for growth continue to be exponential in character.

Currently, the Internet runs across telephone lines, with the packets of information constituting any particular message 'switched' and routed by special computers managed by Internet Service Providers. Voice messages travel along the same trunk lines as data, but voices are switched and routed at separate switches maintained for this purpose by telephone companies.

In the near future, computers and communications capabilities will converge, with telecommunications companies transmitting far more data than voice communications. In addition, the Internet services, once limited to data, will carry voice. Eventually, one system will emerge, and that system that will eventually expand to carry voice, data, and video, To support these developments, a more robust optical-fibre backbone will be installed, eventually linking more seamlessly than today with satellite-to-ground and satellite-to-satellite transmission.[3] As bandwidth expands, real-time video-conferencing will become commonplace, enabling people to 'meet' in cyberspace without having to travel physical distances.

As the information revolution evolves, computers will become easier to use and will appear more 'intelligent' in a human sense. Industry experts expect that, eventually, networking, computing, and other tools of communication will be miniaturized to the point that the tools themselves (but not their function) nearly disappear. The boxes containing computers will become so small that they are nearly invisible; human-computer interaction will be so facile that computers will require almost no skill to use. Intelligence and memory will be embedded in a host of other machines and environments around us, until most products are 'smart' and interconnected.

Within 15 to 20 years, there may be another 'computer revolution' – one that uses the rotations found in all atomic particles to take the place of digital switches. New computers based on quantum computing will use the rotation of atomic particles in the way the '0s' and '1s' of current digital computers conduct computations. The result will be computing power that is blindingly fast and utterly ubiquitous.

Joined with emerging developments in three related areas, advanced silicon-based computers (and eventually quantum computers) will almost seamlessly interact with their environment:

- natural-language processing – using computing software that understands vernacular speech;
- three-dimensional information processing – using sensors and object-oriented software to enable computers to move about within their environment;
- enhanced computer 'intelligence' software – having computers that can understand, reason, and react to their environment.

Advanced computing will also be networked so that intelligent control of the environment is available to almost anyone. Homes will be 'smart' – they will monitor temperature, repair and clean themselves, and order and prepare food or perform other chores at the request of the homeowner. They will learn patterns of use by the homeowner or user and begin to take over simple tasks. The same intelligent design and ease of use will be true of other machines. Once people adapt to ubiquitous computing, they will cease to be aware of it as separate from them and begin to view it as part of the environment, much as we do with pens, electric lights, and eyeglasses, all of which, in their own time, were 'new-fangled' technological breakthroughs.

While ubiquitous, networked computing and electronic communications hold tremendous possibilities for convenience, efficiency, and data capacity, there are also significant potential downsides of these capabilities. The first may be the enhanced capacity for surveillance and a corresponding decrease in personal privacy. There has already been considerable discussion of privacy issues with regard to the use and marketing of personal information stored on computer databases. Advanced computing and networking raises a whole host of more specific potential abuses specific to itself.

Second, there is the potential for increased criminal and terrorist activity. The opportunities for this type of activity are already evident, with money launderers, drug traffickers, hate group, and pornographers being among the most innovative users of electronic communications. Internet commerce is inherently borderless, and makes the collection of taxes and regulation of business activities much harder for national governments to carry out. There is also increased possibility for undermining social order by spreading false or misleading information that is difficult to counteract.

Third, cultural communities may find it difficult to put limits on what their members see and hear. In the United States, this has come up primarily as a matter of controlling pornography, with the US Congress tacking on the Communications Decency Act to the 1996 Telecommunications Deregulation bill, a move that was quickly struck down by the courts. While many people in the IT community remain strongly opposed to any abridgments of free speech, control of pornography and particularly of the access of children to pornography remains an important issue for a large number of Americans. In other countries, concerns go further toward the protection of other aspects of cultural heritage: the French, for example, have tried to impose rules concerning the use of the French language on French internet sites.

Fourth, the Internet can influence civil society and social cohesion. As Ithiel de Sola Pool (1983) once observed, the shift from one-to-many mass communications to one-to-one or many-to-many conversations of the sort fostered by modern computer communications had the effect of fragmenting national dialogues by making them much narrower. That is, the lowest common denominator character of mass media during, say, the middle decades of the twentieth century, ensured that all Americans experienced a common set of cultural experiences, whether the Ed Sullivan Show or Life Magazine and the Saturday Evening Post. Not just electronic communications but all forms of media have moved to much more specialized niche markets, with 500-channel cable TV replacing the three national networks and tens of thousands of on-line discussion groups springing up. This shift toward narrower groups has had a beneficial effect of improving the quality of communications in many cases, but raises the possibility that the civil society as a whole has become increasingly fragmented.[4] This means that, in the end, citizens have fewer and fewer common cultural experiences and points of reference, with possible negative implications for their broader ability to associate and work together as a political community.

Finally, there is the simple overload of too much information and too much communication. At some point, the capacity for communication will come up against the capacity of the human body to absorb and use information – that is, at least until genetic manipulation or computer-based prosthetics offer people the chance to increase memory or process information more quickly.

Human genetic manipulation and bio-informatics

Human genetic research, and its application in generic engineering, will offer three capabilities that are not possible today. It will enable deliberately changing human genes to alter damaged or disease-prone genes; it will enable changing genotypes in a way that is transmissible into succeeding generations; and it will replace or enhance human traits, such as strength or intelligence, beyond what is available in nature.

One milestone toward this end is the completion of the initial phase of mapping human genes. A deeper understanding of human genetics, combined with advances in human biology and organic chemistry, seems likely to change the practice of medicine in revolutionary ways.

Sometimes termed 'molecular medicine', the application of genetics to medicine promises to revolutionize that art by applying knowledge of human genetics to individual patient treatment. Understanding biological changes at the genetic and molecular level promises to shift patient care from a standardized set of practices that treat symptoms of disease to individualized treatment for the underlying cause of disease.

Chemical and biological processes are being researched for their ability to detect specific molecular and cellular events accurately and quickly, with the goal of enabling almost immediate diagnosis and treatment. Molecular medicine seeks to enable biological cures – rather than traditional chemical or surgical cures – for biological problems. Costs of diagnosis and treatment are projected to go down and outcomes will improve as diseases are caught and treated at an earlier stage, with treatment tailored for individual success.

Currently, for example, certain cancer tumors may have as little as a twenty percent response rate to a specific treatment. Low probability treatments are combined in hopes that one will work. Standards of care dictate that the treatment is tried in a kind of buck-shot approach to a cure. Molecular medicine will allow diagnosis of the genetic make-up of the tumour, enabling a targeted treatment with a much higher chance of succeeding. Combined with advances in imaging technology and sensors at the molecular level, medical practitioners will be able to diagnose pre-cancerous tissue changes in a way that will enable early treatment and prevention.

As these changes occur in the practice of medicine, the health care delivery system will need to adjust accordingly. The pharmaceutical and medical device industries should begin to shift as well, conducting more research on genetic treatments and producing fewer mass-marketed drugs.

Molecular medicine will be enabled by the information revolution in two ways: rapid diagnostics will likely be done using computers and information technologies. Computers will be used for complex diagnoses and the Internet will be used to conduct remote consultations between physicians. Secondly, information about the patient's full medical history will be instantly available through a storage system either available embedded in a chip in the patient's body, or it will be accessible from a central stored database. These advancements in recording, manipulating, and transmitting biomedical data – made possible by the information revolution – will offer improvements to health care over the next twenty years in ways that are only vaguely dreamed-of now.

Developments in technology at the convergence of biology and information – such as bio-sensors and smart materials – are poised to produce perhaps the greatest changes, and challenges, to social and defense systems. Technologists suggest that miniature sensors detecting chemical and biological information may soon be available, capable of providing instant feedback on individual or group activities. Joined with large-scale networked computing, this information conceivably will be available to a broad group of users. Advances in software systems may soon offer the capability to detect previously unknown patterns within massive amounts of unrelated data. These developments may, on one hand, increase capabilities in important public goods functions such as environmental monitoring and economic management; on the other hand, the possibilities of using these technologies to create new forms of terror, conflict, and abuse of power appear to be quite significant.

The actual merging of carbon and silicon, starting with the creation of computer prosthetics and moving toward implantation of artificial devices are applications that are being researched now. The creation of 'smart' prosthetics is currently being tested as ways to help handicapped persons develop more normal functioning. In its simplest form, widespread use of this capacity may come in the form of a subcutaneous chip that will carry all of a person's medical information – particularly useful when an unconscious patient arrives at the medical emergency room. The fact that health care will become increasingly information-dependent means that access to this information will be essential for good medical care and for the efficiencies it will bring to the health care system. A second application envisioned for this technology is financial – an implanted ATM or debit card that carries a person's financial resources on it for instant access.

Beyond the simple implantation of silicon-based information that can be accessed by external sources is the possibility offered by true 'cyborgian' technology. Technologists have suggested that the most intriguing of these applications is the ability to implant intelligence or knowledge. Research is underway that explores the limits of bio-memetic reverse engineering to understand neurological processes and whether these can be digitized. It appears to be possible to monitor electronic traffic in the optic nerve, for example, because it is transferred as an electrical signal that can be digitized. If it is possible to digitize what is being seen, it would appear to be possible to digitize an image that one person, or a camera, sees and transfer it to brain cells within a person, or to another person or remove location. Under this scenario, a blind person could 'see' using a camera that transmits digital signals to the brain.

To the extent that human consciousness has a quantum basis, a possibility suggested by current research, it may be possible to imagine enhancing brain capacities by using the emerging principles of quantum computing. This would entail not a convergence of silicon and carbon – with all the attendant problems associated with the body rejecting foreign materials – but of using the brain's matter as the basis for computing. Although most scientists reject the reductionist notion that the brain is a type of computer, physicist Roger Penrose has suggested that quantum computation with objective reduction already occurs in the brain and serves as the basis for consciousness, proffering the possibility that the brain could be programmed to use quantum functioning to increase memory or reasoning power (Penrose, 1997).

While these advances offer significant opportunities for reducing human suffering, they also suggest potential for misuse. As Leon Kass pointed out in a paper prepared for a RAND presentation, the possibilities offered by genetic profiling of individuals raise questions, on one hand for the individual, and on the other for society, in ways that we are not prepared to address.[5] For example, Kass pointed out, 'in many cases, practitioners of prenatal diagnosis refuse to do fetal genetic screening in the absence of a prior commitment from the pregnant woman to abort any afflicted fetus.' Moreover, economic pressures to contain health care costs will almost certainly constrain free choice in the face of certain genetic knowledge. Discrimination in insurance may compel certain kinds of genetic intervention that will change the human genotype for future generations. The opportunity for those with funds to access vastly superior health care will likely increase the discrepancies of care between rich and poor.

Finally, the most troublesome of genetic manipulations is the possibility of manipulating human nature itself. This has come up initially in the technology for cloning human beings, where current developments in science suggest that human closing may be possible, copying either a whole person or parts of a person for replacement of diseased or aged organs. Beyond that lies the prospect for germ-line manipulations that will change the characteristics and behavior not just of individuals, but of all of that individual's subsequent descendants. The moral questions raised by this technology are too numerous and difficult to expand upon in this chapter, but there are numerous possible social misuses of this kind of technology, from widening existing gaps in wealth and life opportunities, to producing unintended consequences by altering important but poorly understood aspects of human behavior.

Nano-science and the challenge to the concepts of work

An understanding of the nano (or atomic level) structure of matter, and ways to manipulate that structure, has exploded in recent years. Due in part to the study of physics at the quantum level, as well as vastly improved laboratory equipment, physicist and now engineers have gained knowledge of how matter is composed. The goal of the emerging field of nano-technology is to build structures starting at the atomic level, either improving upon natural structures (by making ceramics as strong as steel, for example) or by recreating natural structures at less cost. Already, nano-manufacturing has demonstrated that it can change the function of materials as well as the fundamental nature of how materials are produced. Other possibilities include:

- creating biological materials that can sense and transmit information to create a sensor network throughout the body;
- creating drug delivery systems that travel within the bloodstream and can target drug delivery at the atomic and molecular levels;
- building revolutionary super-hard or superconducting materials like friction-free power transmission lines or plastics as hard as steel.

These possibilities clearly suggest a change in the nature of production and labor, perhaps placing manufacturing capability in the hands of the individual, where custom products can be created from the atomic level in a person's home. The very nature of 'labor' is about to undergo a fundamental transformation, and, with it, social structures will change just as dramatically. The implications of this for our market economy and distribution of wealth are unprecedented.

Karl Marx noted in *Capital* that capital-goods industries benefit from the accumulation of capital, making industrialization and the factory system possible. Technologies of production in the factory system moved control out of the hands of the workers; Marx's analysis led to a social movement that 'raged against the machine'. Nano-science (and similar changes promised by information and biological technologies) will move technology radically back into the hands of the individual worker and consumer. The collective decisions of many individuals will certainly affect the social order, but the 'rage' will have to be aimed at the individual rather than the machine.

The problem of governance

This section explores the particular governance challenges emerging from the dual revolutions, with a focus on the issue of domain names on the Internet and biotechnology issues. The section immediately below defines governance and describes why these technologies offer particular governance challenges to the international community.

What is governance?

As John de la Mothe has noted in Chapter 1, governance is the effort of human communities to try to control, direct, shape, or regulate certain kinds of activities. At one extreme, these communities can try to ban the activity altogether. Governance is almost always carried out by states, although it can be carried out in other ways as well. In continental Europe, for example, there is a corporatist tradition in which the state works with certain designated representatives of civil society like labor unions or employer federations to achieve public ends. In the United States, the government has tried to offload many governance functions onto either the private sector, or else onto non-governmental organizations (NGOs).[6]

Governance usually constitutes a subset of public policy more broadly. Particularly in the field of science and technology, there are many types of public policy interventions that are more limited in scope, seeking, for example, to stimulate the development of a particular technology, to promote information-sharing, or to fix certain specific problems. Governance usually involves an ongoing process of regulation and control, and therefore the establishment of institutions designed to effect that control.

It should be clear from the outset that any effort to create governance institutions for either of the technology areas in question must be international in scope. Modern information technology in particular is inherently borderless: the network does not care where the physical location of a server is; an effort by one nation to control or close down a site can be defeated by moving it to another country or jurisdiction. Indeed, the mobility of modern information has led some observers to argue that even a comprehensive international governance regime would be incapable of taxing, regulating, or otherwise controlling the flow of data and information. Biotechnology is less mobile, but still presents many of the same challenges: if one country wants to ban, for example, cloning or designer babies, people who want such things can simply get on an aeroplane and obtain them in another country without such regulations. The materials needed to manufacture products from the atomic level up will be easily available anywhere in the world. It is useless, therefore, to think about governance except in an international context.

Changing attitudes toward governance

It is clear that the last 50 years of the twentieth century have seen massive changes in attitudes toward the workability and desirability of governance, and nowhere more so than in the United States. While the Great Depression and World War II ushered in an era of increasing government intervention in many aspects of economic and social life and the creation of the modern welfare state, the latter part of the century has reversed these trends and emphasized privatization, deregulation, downsizing of bureaucracy, and private, market-based solutions to many social problems.

This is nowhere more evident than in the field of telecommunications and IT more broadly. In the first half of the twentieth century, virtually all countries that built telecommunications infrastructures did so with state-owned PTTs (post, telephones, and telecommunication organizations) that exercised monopoly control over their national systems; the United States used a private firm, AT&T, but heavily regulated it in the interest of the Federal Communications Commission.[7] Traditional PTTs were highly centralized, bureaucratic, and exercised dictatorial control over technology development and implementation in their respective national jurisdictions. They saw fit to set prices for services and equipment, to mandate universal service and a system of cross-subsidies, and were the sole representatives of the users of telecommunications services in any kind of international negotiations.

The international governance mechanisms that were set up in this period were similarly centralized and top-down. The International Telecommunications Union (ITU) in Geneva was simply the supranational regulatory body established by national regulatory bodies through international agreement. Like the PTTs that were its clients, it too sought to set a single, authoritative set of rules regarding frequency allocations, tariffs, protocols, and the like.

It is safe to say that very few advanced countries would set up a similar governance system on either a national or international level if they had to build one from scratch today. Most countries have privatized their PTTs (like the sale of British Telecom or Deutsche Telekom) or have deregulated their telecommunications sector (as in the case of the 1984 consent decree that broke up AT&T, or the 1996 Telecommunications Deregulation Act). Countries are much more likely to favor private firms over state-owned monopolies to provide telecommunications and IT services, and to accept a higher degree of competition in the provision of these services.

The reasons for this shift are both normative and practical. On normative grounds the modern, centralized welfare state was seen by the 1980s to be at a dead end, stifling creativity and innovation while wasting taxpayers' money through mismanagement and inefficiency. Indeed, the central ideological divide has focussed on the role of the state in regulating private markets and providing social welfare. In this regard, the United States has taken the lead, being typically exceptional by pushing out the envelope of liberalization.[8]

There is a further normative reason for wanting to minimize the regulation of IT that is specific to this type of technology. IT is widely perceived as being supportive of the sorts of political values that are emphasized in the West. Cheap, ubiquitous information technology from phones, faxes, and radios to computers, e-mail, and the Internet, have been seen as promoting a deconcentration of power throughout the world and promoting the spread of liberal democracy.[9] Modern communications was seen as critical in undermining the legitimacy of communist states like the former East Germany and the Soviet Union (Shane, 1994), as well as the right-wing dictatorship of Ferdinand Marcos in the Phillippines (Ganley, 1991). In the future, it is widely believed, IT will help open up closed societies like the People's Republic of China and Singapore, whose governments have tried to control Internet use for political reasons.

Americans, however, have special reasons for wanting IT to spread around the world with as few constraints as possible. Americans stand to benefit from the IT revolution in power and economic terms. It is American companies that dominate the global IT industry and are on the cutting edge of innovation there; it is American cultural products and media, from CNN to Disney to MTV, that will be carried by this equipment; and it is American values, both political and cultural, that will be fostered as the world grows more electronically connected. It is natural that

Americans should argue in favor of a minimal IT governance regime since if favors both their ideology and their self-interest.

On a practical level, it became clear that the evolution of technology itself was making the old governance model inappropriate and even counterproductive. Governments established PTTs in the belief that a telecommunications network constituted a natural monopoly, and that private markets could not be relied upon to run such a network in the public interest. The advance of technology, and in particular the advent of long-distance microwave transmission, convinced regulators in the United States by the 1980s that competition in long-distance service was feasible. By the 1990s, other innovations like cable provided alternative routes in the local loop, and led the US Congress to permit a much more far-reaching deregulation. (Huber and Kellogg, 1994.) It became clear that PTTs and national regulatory bodies moved much too slowly to provide useful standards and rules in this fast-changing area. What applied on a national level was doubly true on an international one for a body like the ITU. By the time the latter could agree on, for example, a communications protocol, technology had moved on and made the new protocol obsolete. These regulatory institutions were seen as positive hindrances to technological advance.

What was true of telecommunications was even more true of computers, software, networking, and other aspects of information technology (IT). Computer technology grew up in an almost completely unregulated environment. IT did not particularly reward scale; in some parts of the industry, at some points in the development cycle it has been better to be a small and nimble competitor than a large, hierarchical corporation, so the spectre of concentrated power that had driven regulation in other sectors did not exist in so obvious a form. The US Justice Department's anti-trust suits against IBM and AT&T in the 1970s and 1980s were widely seen as counterproductive given the fact that both corporations were soon upstaged by smaller rivals. The underlying rate of technological innovation and change in the IT business was so rapid that the government was loathe to intervene, though it finally did so in cases where private firms like Microsoft and Intel developed quasi-monopolies in PC software and hardware. The government's pursuit of these companies was, however, controversial, and virtually no one advocated encumbering the IT industry in a regulatory framework like the one that surrounded telecommunications.

What is true on a national level is doubly so on an international one. The ITU has by and large given up trying to set standards for the IT industry; such standards that exist come out of private industry consortia like the Institute of Electrical and Electronic Engineers (IEEE) or the Internet Engineering Task Force (IETF). These bodies are less hierarchical than the ITU; participation is voluntary and based on the desire of market participants to have a common set of standards. (It is true that the Internet itself and the use of the TCP/IP protocol for Internet communications was due to an act of government intervention in the form of the US Department of Defense's creation of ARPAnet; but once the Internet was privatized, further growth and technological innovation took place rapidly and in a highly decentralized form.)

The fact that IT regulation is difficult and likely to be counterproductive has led many observers to argue that governance is not desirable and should not be attempted. Attempts to tax electronic commerce or the Internet more generally have met with particular resistance in the United States, and with a great deal of special pleading that alternates between arguments that it is illegitimate to tax electronic commerce to arguments that it is technically impossible to accomplish.

Internet domain name regulation

The fact is, however, that IT poses a number of governance issues that call for the creation of new institutions. One clear case is the assignment of Internet domain names. Domain names (that is, Internet addresses ending in .com, .gov, .edu, etc.) have become necessary to electronic commerce and are commonly seen as extensions of corporate trademarks. As such, they constitute a form of property, and require a mechanism for property rights protection. A company like McDonald's or Coca-Cola that has invested millions of advertising dollars in building up a brand name will not take kindly to another firm or individual registering mcdonalds.com or coke.com on the Internet.

In the early days of the Internet, domain names were assigned by the US government. When ARPAnet was turned over to the National Science Foundation to administer and then privatized in 1990, it signed a contract with a private contractor, Network Solutions, Inc. (NSI), to run the administration center InterNIC to provide domain name registration services.[10] NSI simply assigned names on a first-come, first-served basis for a nominal fee. The value of certain domain names quickly created a secondary market for them, however, and new phenomena like cybersquatting in which an individual would register large numbers of domain names in the hope of being able to charge the legitimate trademark owners for them. NSI had to navigate its way through a number of cultural and international issues. For example, while it tried to avoid assigning indecent words as domain names, it could not guarantee that the names it assigned would not be culturally offensive in any of the languages used on the Internet. There is a further issue, moreover, that countries other than the United States require country-specific top-level domains (e.g. UK, France, Denmark), which is regarded by some as a mark of second-class citizenship.

The international assignment of domain names should in principle be no different than the international registering of trademarks, patents, and other forms of intellectual property.[11] In the latter case, there are formal governmental institutions and international legal agreements established for their registration and protection, i.e., a formal system of international governance. If the Internet had somehow been invented fifty years earlier, when government regulation was in greater vogue, it would seem likely that the process of Internet domain name registration would have been turned over to a formal, hierarchical national regulatory organization in the United States, and to an ITU-like body (if not the ITU itself) on an international level. Given that this issue came up in the 1990s, however, it was solved in a typically less statist form. In July 1997, the Clinton Administration ordered the Commerce Department to privatize the process of domain name registration and created a private, not-for-profit organization called the Internet Corporation for Assigned Names and Numbers (ICANN) to oversee the process. ICANN, in turn, took away NSI's monopoly in domain name registration and proposed to turn the task over to five private companies that had passed its accreditation standard.

The ICANN represents a strange and in some ways unprecedented hybrid form of governance. It is not a formal government organization, nor is it a purely private sector group like the IEEE. It seeks to build consensus through informal methods, without enforcement powers and without the legitimacy that would come about were it established as the result of formal inter-governmental agreement. The hope on the part of its designers was evidently to retain the informality and flexibility of the kinds of governance institutions that characterized the Internet itself in its early days, while meeting the needs of the Internet's increasingly heterogeneous users.

The question, however, is whether domain name regulation can be adequately governed by this type of informal body. Informal coordination mechanisms require a high degree of consensus over goals and methods. NSI has, naturally, opposed the dilution of its monopoly, and has refused to recognize ICANN's authority. There are broader questions, moreover, within the Internet community as to the rules by which ICANN board members were selected (Zangardi, 1999: 52). One can easily imagine these disputes growing more severe in ways that will throw this entire governance system into jeopardy. It is not clear why domain name registration should be limited to five companies, what the grounds were for accreditation, and how larger international disputes are to be settled. Supposing, for example, that a powerful nation like China rejects ICANN's authority or complains about the way that it has assigned names to Taiwanese firms.[12] When such conflicts become too severe, they rapidly outgrow the capability of informal mechanisms and get kicked upwards to an inter-governmental level.

The need for more formal governance institutions may become evident even in fairly technical areas like the setting of standards and protocols. The IETF, unlike a protocol-setting body within a traditional telephone company, is famous for being decentralized and bottom-up. Ideas filter in from all over the Internet community, and if they are workable, they are simply ratified by the IETF rather than mandated out of the blue. Some people have described this as a Darwinian, evolutionary process which fosters rapid innovation. The problem, as in the case of domain name regulation, is that such solutions oftentimes do not scale well. The number of participants in IETF meetings tripled between 1987 and 1989 as the number of stakeholders increased (Zangardi, 1999: 17). A body with a more diverse membership and conflicting commercial interests is likely to have much more severe difficulties reaching consensus than the more inbred, homogeneous one that characterized the Internet in its early days.

IT and the development of the Internet has therefore not ended the need for governance, but rather has raised new issues (like domain name regulation) that call for institutional solutions. There is, none the less, a presumption that whatever regulatory mechanisms are ultimately established ought to be limited in scope, flexible, and based on broad consensus rather that hierarchical fiat. The reasons for this exist not only on a practical technological and economic plane, but also on a normative one: very few people perceive the worldwide spread of information technology to be a threat to important social values.

Biotechnology

The same cannot be said, however, for biotechnology, which is viewed with much greater alarm by publics throughout the developed world. Genetically altered foods have already been in the marketplace for several years, and have been regulated in the United States by the Food and Drug Administration from the start because of potential safety concerns. The level of regulation is considerably higher in Europe, which has sought to limit, control, and in some cases ban outright certain kinds of biotechnology. European efforts to restrict imports of American biotechnology products, particularly the EU's ban on beef that has been treated with genetically altered hormones, have led to severe trade disputes between the US and the European Union.

The reasons for this resistance to biotechnology are complex. In Europe, it is often related to memories of Nazism and the genetic experimentation that went on in the National Socialist period. Europe has developed a very powerful green movement in the past generation, a movement that for ideological reasons does not like tampering with nature. There are also strong concerns over the safety of genetically altered plants and animals, fears that have been stoked in the popular press by stories about 'monster tomatoes' and the like.

The contrast between European and American attitudes toward biotechnology is striking and has led to important misunderstandings on both sides of the Atlantic. Americans tend to be eternally optimistic about technology and its social impact, and regard those raising concerns about biotech as Luddites who are frightened of change. Ethical concerns about genetic engineering tend to focus on the narrow issue of safety, and whether regulatory and testing procedures are adequate to permit the commercialization of biotech products. The commission appointed by President Clinton to look into the ethical aspects of human cloning argued for a temporary ban, but only on the grounds that the procedure could not be safely carried out. The presumption was that once the safety questions had been answered, there would be no further objections to individuals proceeding to clone themselves. In trade negotiations, Americans tend to focus on safety issues. When the science on this issue is on their side (as it is in the case of beef hormones), they tend to interpret European motives as either dishonest, ignorant, or representing a covert form of protectionism.

As Leon Kass pointed out in the paper he wrote for a RAND study group (Fukuyama, 1999), there are, however, any number of reasons why societies might want to decide to restrict the use of biotechnology, that have nothing to do with safety. Many people have religious or ethical reasons for not wanting to manipulate nature on a genetic level. When biotech advocates point out that we already alter gene pools in numerous ways (for example, through simple cross-breeding of plant and animal species), biotech skeptics suggest that we are on a slippery slope to ever-greater interventions that may some day alter some of the most basic characteristics of the natural world. The most important of these is human nature itself. As discussed above, so-called 'germline' research of the future will differ from medical technology of the past in its potential to alter human nature by affecting not just the individual to whom it is applied, but all subsequent descendants of that individual. We may one day be able to have not just simple clones of ourselves, but human beings whose behavioral characteristics we have in critical respects been able to create ourselves.

Many of the proponents of biotechnology will argue that this kind of observation is unduly dramatic and alarmist. The purpose of research in biotechnology is therapeutic: it aims at uncovering what are now clearly understood to be the genetic underpinnings of diseases like breast cancer, Alzheimer's, schizophrenia, and the like, and to provide cures for them. Germline research, it can be argued, simply takes this form of therapy to its logical conclusion: if the propensity for a disease lies in a genetically heritable characteristic, what is wrong in principle with a genetic intervention designed to eliminate that propensity from present and all future generations that might suffer from it?

The fact that there is no clear answer to the last question suggests, as Leon Kass has pointed out, why biotechnology will be so hard to resist in the future: any potential negative consequences of genetic manipulation will be intimately connected with positive benefits that will be obvious and measurable. Many people argue that we can draw a line between therapy and enhancement, and that we can reserve genetic engineering for the former. But drawing boundaries in grey areas will be much easier said than done. There is general consensus that some conditions like schizophrenia are pathological; the problem is that there is no consensus as to what constitutes health. If one can administer growth hormone to a child suffering from dwarfism, why not to one who is in the fifth percentile for height? And if it is legitimate to give it to a child in the fifth percentile, then why not to a child in the fiftieth, who wants to receive the clear-cut benefits of tallness?

All of this suggests that societies will want to control biotech to a much greater degree than they have wanted to control information technology, but that establishing limits, norms, and principles will be extraordinarily difficult. The same problem of implementation arises in biotech much the same way as it does with IT: assuming that societies can agree on limits to the technology (e.g., cloning or designer babies), how can they make these limits stick in an increasingly globalized world?

Governance of the revolutions

As we have seen, the IT revolution has had beneficial effects in undermining authoritarian hierarchies and distributing power more broadly. In popular imagination, IT is seen as good for democracy, good for the economy, and (if they are Americans) good for the United States as well, since it is the US that dominates the global IT industry. Biotech, on the other hand, while having unquestionably beneficial effects, is regarded by many non-scientists with much greater suspicion.

Efforts to control biotechnology will run into the same practical hurdles as attempts to control IT. Globalization means that any sovereign state seeking to impose limits to, say, cloning or the creation of designed babies will not be able to do so; couples facing a ban imposed by the US Congress, for example, may be able to slip into the Cayman Islands or Mexico to have their cloned child. Moreover, international competition may induce nations to cast aside their qualms: if one country or region of the world appears to be producing genetically superior individuals through its relaxed rules on biotechnology, there will be pressure for other countries to catch up. The libertarian mindset and the absence of international governance mechanisms, that seemed appropriate for the largely benevolent IT revolution, may be less appropriate for a more sinister biotech revolution. But at that point, efforts to close the gate may be unavailing.

The revolutions may have an interactive effect on one another. The belief that technological innovation is necessarily beneficial is not one that has been universally shared, even by Americans. The advent of nuclear weapons, after all, was greeted with great alarm and led to the creation of a vast international regime to prevent the technology from spreading. But the largely positive and beneficial nature of IT, coupled with the anti-statist attitudes of the late twentieth century, have shifted attitudes toward technology more generally and disinclined many people from even considering regulatory schemes. When a new technology like germline engineering or cloning comes along that societies do want to control, we find ourselves in a situation where we are bereft of international institutions that are capable of imposing rules on a global basis. What international institutions we do have, like the World Trade Organization, are biased against the regulation of technology because it tends to see such constraints as non-tariff barriers to trade.

And indeed, the practical obstacles to establishing a formal, hierarchical governance system are tremendous. The world has changed enormously since an organization like the ITU was established in 1865. The number of actors on the global scene has more than doubled, and the level of technological sophistication and economic development outside of the club of Western powers has vastly increased. Any international regulatory scheme, not just of IT or biotech, but of child labor, occupational health and safety, or environment, would have to be hammered out by a group of culturally diverse countries with vastly different levels of development and economic interests. Fifty years ago, for example, no international conference would have regarded the views of China as significant; today, no international agreement can go forward without its approval.

Possible government actions

Because the technologies emerging from the information and biological revolutions are inherently global, success in governing these technologies depends upon the ability to enlist all stakeholders – states, NGOs, interest organizations, and citizens – to cooperate in support of the development of governance norms or structures. A number of interests intermingle in the applications and governance of these technologies – commercial, defense, social, individual – and these various interests can be made to work to the advantage of cooperation.

As RAND researchers, John Arquilla and David Ronfeldt, point out, in the economic-legal sphere of governance of new technologies, the primary concerns are commercial.[13] Given the overriding interests of many parties in ensuring safety, efficiency, and security of the flows of information and goods that can be traded as a result of electronic commerce, cooperation may depend upon reaching agreement in 'substantive law' – agreeing on what constitutes criminal activity on the Internet.

Cooperation may also hinge upon acceptance of a body of administrative and legal procedures that would establish jurisdiction and allow enforcement of substantive laws designed to protect property and other assets, both in and out of cyberspace. In the information realm, agreement about such matters as territoriality, extradition, and the notion of 'hot pursuit' may form a minimum basis for international cooperation – especially in the area of cyberspace-based territoriality,[14] etc.

Law can be decided on a cooperative basis and implemented by a top-down approach. Norms governing non-criminal use of the Internet may actually emerge from practice: Huberman's research has suggested that a number of these governing functions will emerge from current practice, leading to self-regulation of the Internet without having to institute a governing agent.

In the social arena, primary concerns are more numerous and varied than those in the commercial realm, making global cooperation more difficult to achieve. The complexity and speed of technological change in information and biotechnology suggests that top-down approaches to governance will not work. Given the difficulty of achieving global consensus on governance in areas such as genetic manipulation, trade in body parts, cloning, implantation of silicon chips, enhanced intelligence and other emerging technologies, it may be worth sorting out which of these issues are worth attempting to address, if not directly with regulation, then in a global dialogue structure to monitor and oversee their use and combat misuse.

Top-down or positivist approaches to governance of these technologies will not work. Moreover, the individualistic nature of the control of these technologies and their applications and use make governance particularly difficult. Given that many decisions about use and application will be made on an individual basis, it is hard to imagine any governance structure without wide 'buy-in' from the polity. Accordingly, one approach to making decisions like these might be to use a distributed decision-making model that would involve a significant number of citizens in deciding what technologies to support with research and development funds, what technologies need governance, what norms of use and application there should be, and if, how and at what level of formality technologies should be regulated.

Broader decision-making models

A possible approach to involving a broad swath of the polity in decision-making about technology would be the creation of citizen councils formed to provide input to higher-level, more formal governing bodies. One model might involve aiding the organization of hundreds of citizen councils across the US (or even around the world), and encourage deliberation about norms of use, regulation, and governance of technology. Using the networking capacities of information technology, such councils could conceivably deliberate and share input on a series of governance questions in a way that draws toward a consensus of views on how to manage and govern technologies.

Councils such as these would have the feature of allowing norms to emerge and develop at the level of the user. As we have stated, in cases where many individual decisions will lead to collective action, norms may be preferable to formal governance. Norms are more likely to emerge and be acted upon at the lowest possible level of decision-making – thus the councils could help to create and distribute these ideas. In addition, the councils would also serve an educating function, helping to improve the public understanding of science – an area that many policy-makers see wanting in the United States.

Citizen councils like this have actually been used quite effectively in Europe. The mechanism of 'consensus councils' has a long tradition of settling contentious matters in science. These councils, however, have traditionally been made up of recognized experts and professionals in the field to be considered, leading one observer to liken them to a 'synod of Bishops' (Skrabenek, 1990). In the late 1980s, however, the Danish Board of Technology redefined consensus councils as bodies of lay citizens that would be convened to consider the evidence on a particular science or technology issue, participate in public debates, and ultimately provide a consensus report of their findings and policy recommendations. The purpose of the process was not to dictate policy but instead to help the legislature understand where an educated population might stand on an issue before considering specific policies. The consensus council reports have not only shaped policies; industries have also used them to craft research agendas to avoid public opposition that might emerge after they have made significant investments, when it is difficult to change direction (Sclove, 1996: 24—8). This success led to the engagement of similar processes in the Netherlands, the United Kingdom, and Australia (Schibeci *et al.*, 1999; Sclove, 1996). A similar citizens' panel was convened in April 1997 at Tufts University in the United States to consider the topic of 'Telecommunications and the Future of Democracy' (Hackman, 1997: 6—7).

It may be necessary to determine the proper scale at which this sort of public discussion and consensus might occur, particularly in a country as large and as populous as the US. Similarly, it would be difficult for fifteen people to approximate the demographics of this country. Also, finding a body viewed as non-partisan may be difficult in the US context. Still it seems that this sort of considered lay opinion has clear advantages for governance that can complement mechanisms such as opinion polls and expert panels and reports.

Governance by NGO

Another model for future governance is the action of non-governmental organizations (NGOs) (Matthews, 1997: 50–66). There are numerous recent examples where NGOs, empowered by low-cost electronic communications, have been able to act to bring about outcomes that could not be achieved by sovereign nation states, either acting alone or in concert with one another. One case of this was the intervention by human rights and other activist groups on behalf of the Indians in Chiapas, Mexico, whose rapid action forced the hand of the Mexican government in that instance.[15] More recently, Greenpeace and other international environmentalist groups were able to force Shell Oil to change policy both in Brent Spar in the North Sea and with respect to Nigeria. Other advocacy groups forced the sportswear maker Nike to make promises with regard to the use of child labor. Environmental groups have also been active on a number of biotechnology issues, such as the dispute over beef hormones.

In each of these cases, the NGOs in question were able to change the behavior of a large multinational corporation, or in the Chiapas case, the government of Mexico, in a situation where state action was ineffective. There have, of course, been questions raised as to the true effectiveness of this kind of intervention: who will monitor Nike, for example, for compliance with the promises it made on labor standards? None the less, NGOs have the capability of organizing quickly and transnationally in ways that avoid the bureaucracy and rigidity of conventional international organizations.

There are a number of problems with this model of governance, however. The first concerns legitimacy. One of the reasons that formal government institutions exist is to confer legitimacy on their decisions; in an age when legitimacy results from popular consent, democratic institutions like legislatures and parliaments are broadly accepted as a means of expressing consensus. Government by NGO, on the other hand, comes about simply as a result of groups being able to organize themselves effectively, with no requirements for being representative or promoting the public interest. Many existing NGOs claim to represent the public interest, of course, and many see themselves as offsetting the inordinate power of multinational corporations or governments in the pay of multinationals. But they are self-appointed tribunes of the public interest, whose actions may or may not correspond to what democratic publics really want. They are also unaccountable; unlike a democratically elected legislature that can be turned out of office in the next electoral cycle, NGOs cannot be removed by popular demand.

A further problem is that the kinds of issues that NGOs can deal with effectively are limited in many ways. Nike and Shell could be pressured by activist organizations only because they had large consumer marketing operations that could be hurt by bad publicity. Companies that sell to other businesses, or which market unique or difficult-to-obtain products, would be far less vulnerable to pressure, and indeed much less open to public scrutiny.

Conclusion

Today's governance structures are challenged by a unique shift from collective control and hierarchical decision-making to individual control and decision-making that will mark the technologies emerging from the revolutions in information, biotechnology, and nano-science. The very nature of these technologies makes regulating and controlling them particularly challenging. Traditional 'top-down' or positivist methods of governance will have little influence over how these technologies are developed, diffused, and assimilated. New governance mechanisms are needed, and they must emerge quickly, be flexible, and have broad buy-in. The alternative methods discussed here – standard-setting bodies, citizen councils, and NGOs – present some options to policy-makers considering ways to deal with these challenges, but only begin to suggest the challenges ahead as the technologies change social order.

Notes

1. See both Thompson, E. P. (1996) and Kirkpatrick Sale (1996).
2. As an example, in order to fully utilize technologies in the classroom, educational reform measures must be put in place. Teachers need to change the way they offer instruction and organize the classes. This change will come about slowly, and, then, only if it proves to increase learning in the classroom. See 'Fostering the use of educational technology: elements of a national strategy', Thomas K. Glennan and Arthur Melmed, Santa Monica, CA: RAND Corporation, MR-682-OSTP/ED, 1996.
3. Changes expected from these technologies are discussed in more detail in 'New forces at work: industry views critical technologies', Steven Popper, Caroline Wagner, and Eric Larson,

Santa Monica, CA: RAND Corporation, MR-1008-OSTP, 1998.

4. For a broader discussion of this issue, see Fukuyama (1999: ch.4).

5. Francis Fukuyama and Caroline Wagner, 'Information and biological revolutions: global governance challenge – summary of a Study Group', MR-1139-DARPA, Santa Monica, CA: RAND, 1999.

6. For an overview of the role of 'third sector' organizations in modern government, see Salamon (1995).

7. See Noam, E. (1992).

8. On American exceptionalism with regard to anti-statism and market liberalism, see Lipset, S. M. (1995).

9. On this point, see Peter Huber (1994) and Thomas L. Friedman (1999).

10. See John A. Zangardi, 'Regulation of the Internet Domain Name System', unpublished paper, 1999.

11. With the difference that Internet domain names are inherently global. That is, it is quite conceivable that the name McDonald's can be simultaneously registered as a trademark to a US fast food chain in the US and to an engineering firm in Scotland; by contrast, there can be only one mcdonalds.com.

12. A group in the People's Republic of China has actually registered the domain name 'taiwan.com,' much to the annoyance of Taipei.

13. John Arquilla and David Ronfeldt, 'The Emergence of Noopolitik: Toward an American Information Strategy (Santa Monica, CA: RAND Corporation, MR-1033-OSD, 1999), p. 57.

14. See note 13.

15. David Ronfeldt and Cathryn L. Thorup, 'North America in the Era of Citizen Networks: State, Society, and Security' (Santa Monica, CA: RAND Corporation, DRU-459-RC/FF, 1993).

CHAPTER THIRTEEN

Science, Democracy and Technological Risks

Jean-Jacques Salomon

Introduction

Theme which I propose to deal with in this chapter is not an easy one since it illustrates the combination of several factors, institutions and issues which are today at the heart of our political and governance processes. As articulated already in chapters by John de la Mothe, Sanford Lakoff and Michael Gibbons, we have entered a new era in the relationship between knowledge and power, an era where the Cartesian rule of methodological doubt needs to be applied to decisions in a context of uncertainty. In short, the question I want to raise is the following: 'can the practice of such a doubt remain the monopoly of experts?'

Freedom and security

The trade-off between freedom and security has been the subject of debate ever since the very beginnings of political thinking about democracy. Indeed, in democratic societies security is considered to be a right, and the State is held largely responsible for ensuring it. This is not a prerogative just like any other. The modern State developed in the seventeenth and eighteenth centuries on the basis of a 'contract' that explicitly linked recognition of the government's legitimacy – and public acceptance of taxation – to a guarantee that the citizens would be protected. The State is permitted a monopoly of legitimized violence on two conditions: that limits are set to the use of such violence, and that the State must take full responsibility for protecting individuals and property. Michel Foucault has defined the ideal government as one that 'governs close enough to the dividing line between too much and too little security'.[1]

Neither too much nor too little: the ideal balance is close enough to the dividing line between one extreme and the other, in the grey area which would be considered quite reasonable if everyone were in agreement. But what happens where there is no consensus? It is even harder to achieve a compromise because we are not dealing here with purely rational attitudes. And the contradiction is particularly obvious in democratic societies because they are the ones in which demands for greater individual freedom are combined with demands for increased security. Tocqueville noted this tendency back in the nineteenth century, and the modern world, with its far greater complexity and heightened awareness of risks to security, still constantly provides examples of the tensions and crises that arise from this contradiction.

If security were given absolute priority over all other considerations it would mean either imposing totalitarian rule or complete paralysis: either a police state and a military solution to the problem of risk, or else a rejection of innovation and a timed, unchanging society. Democracies do not refuse only to face this dilemma. They insist that the rules of the game and the decisions relating to the trade-off between freedom and security are not settled behind closed doors, in the secrecy of what C. P. Snow (1961) called 'the corridors of power', where decision-makers encounter only technicians and where experts increasingly have the upper hand. Technocratic rule is usually understood to mean that power is taken away from the representatives of the people. The demand for transparency thus tends to challenge a style of risk management that reduces the complex task of governing human beings to no more than a matter of administering things.

From some angles the demand for transparency seems to be outrageous or even unreal: is it possible to know everything and, what is more, monitor every step in a chain of decision-making? It also means dealing with the risk that decisions will be taken without the people or their representatives having any right to be involved. But lack of transparency means breaking the contract. If governments are to have legitimacy and also if the public is to accept taxation – the two things that underpin the guarantee of security for the governed – then public opinion must be kept informed, i.e. the technical establishment must not withhold information. In areas such as waste disposal or environmental protection, the example of what not to do offered by the totalitarian regimes over the last 50 years is so much worse than anything we could have imagined or guessed that one can only agree with Churchill's remark that democracy is the worst possible system ... except for all the rest. I would go so far as to say that in totalitarian regimes the disregard for industrial or technological risks was on a par with their violent treatment of human beings: the two went together.

There are circumstances within democratic societies where decisions are deliberately taken in the dark or in secret, but they are by definition exceptional: war, terrorism, natural disasters. Nevertheless, even in extreme cases, although transparency may be restricted, it is never completely absent. Because it is characteristic of democratic societies that they always demand that decisions be publicized and discussed, even in the most exceptional circumstances – that is perhaps their weakness and also their strength.

Transparency

The notion of transparency was applied to institutions first in the United States, at the time when authority and the 'establishment' were being challenged in the late 1960s. The movement led, among other things, to the 'Freedom of Information Act', which allows individuals access to their confidential files held by public institutions. An 'open' institution is one that hides nothing. Therein lies the paradox of the demand for transparency: its target is what is supposed to be public, but is not. Any suspicion of the contrary gives rise to complaints what information is being withheld, and to fears of manipulation or even machinations by the institution, which has no intention of sharing its knowledge with outsiders.

In the United States, the way the concept evolved in recent years has as much to do with the Vietnam War as with Watergate. The 'credibility gap' with regard to public institutions suggests that – for all that the country is a democracy – its institutions are manipulated and manipulable just as in a society without the same supposed safeguards. In fact, well before the era of major challenges, sociologists like Packard and Mills were criticizing the practices of both public and private institutions that deliberately prevented anyone from seeing precisely what was going on. John Kenneth Galbraith kept up the attack in *The New Industrial State*. A veil is drawn across the truth and the power struggles that take place hidden from the public gaze. There is definitely a link between the increasingly strong concerns about the environment and the demands for greater transparency, which may raise a smile at the Rousseau-like inspiration: the notion of transparency is connected with the notion of original purity, yet surely this is the exact opposite of the industrialized world in which human beings must now live? Lack of transparency in human institutions is thus set against the limpid clarity of nature. But does that really exist?

More seriously, the demand for transparency arises from two phenomena. Both are connected with the increasingly scientific nature of modern industrial societies. First, technical systems have become so complex that they have to be operated by experts. Secondly, people feel that they are being excluded from any democratic oversight of these activities. The result is the paradox that although these societies have the means of acquiring, processing and sharing information which may give individuals far greater knowledge of the world they live in, this huge growth in the amount of information available is not matched by a similar degree of transparency of society vis-à-vis itself.

On the contrary, the more the media provide instant coverage of world events as they happen and call on 'experts' for comments, the more the information creates a vast blur, if not outright faking of 'events', as we could all see for ourselves during the Gulf War. The use of experts – turned into technocrats, strategists, spin doctors, prophets or gurus – is sometimes the best way of getting round conflicts and muddling the debate. And while the public may be more eager to be informed, as levels of education rise and the middle class grows, public suspicion is far from being reduced when institutions with specialized knowledge – as we have seen in the case of nuclear power – make efforts to anticipate the public's questions. (Kourislky, 1998.)

No institution likes to acknowledge these areas of vulnerability. To issue warnings of dangers is tantamount to admitting weakness. So how much information can be divulged before it starts to have adverse effects? The mere act of seeking information from various public or private institutions is often unacceptable, and can result in misunderstandings and conflicts. However, the limits to the ideal of transparency are set not only by producers of knowledge and information, but also by the public's ability to absorb that information. If the object is to elicit a mature opinion, then the appropriate means of information and education must be mobilized more vigorously and more objectively so as to increase public awareness of risk. Instead of avoiding conflict, this educative effort must acknowledge and deal with the deep feelings, interests and tensions that are an integral part of any industrial enterprise. The key aim is not to limit entrepreneurial freedom but to prevent catastrophes – or at least to reduce their frequency and costs to the individual and society.

You now have the essence of what I wish to stress. Actually, speaking in this framework, I will insist on the fact that we are now far from what was the credo of the first scientific institutions in the very beginning of modern scientific research, namely that the pursuit of knowledge is to be dissociated from politics or even values, as Chapter 2 in this volume so clearly enunciates. Let me just return to the part of a book I published some years ago which dealt with the notion of public participation (Salomon, 1993).

Behind the work there is more often that not a façade or a ritual, and this is exactly what puts our democratic systems under suspicion. The demand for participation is not limited to scientific and technological questions, but also concerns many other less newsworthy issues. But if the hallmark of participation is to voice specific interests in the name of the general interest, certain technological developments give more reasons for mobilizing opinion: they are met everywhere, their repercussions concern everybody and often every society, and public opinion is particularly alive to them because their implications transcend the local interest which were the first to be aroused. Moreover, they worry people all the more when they take the form of complicated and even abstruse technical programs because the specialists and decision-makers concerned are suspected of holding back information.

Participation

Whatever the case, the concept of public participation implies that the action of any techno-structure can be exposed – and indeed, often, opened up – to inspection by individual citizens or concerned associations. Yet, if participation begins with this right of inspection, it can also stop there without satisfying the revealed need for a more democratic control of the decision-making process. As Sherry A. Arnstein (1969: 216–24), a political scientist, has shown long ago, there exist various degrees of participation to which decision-makers consent and hence different levels of control. Arnstein was in fact dealing with participation in decisions not on scientific and technical matters, but on questions affecting underprivileged American minorities in local government – urban renewal, anti-poverty programs and the so-called Model Cities. But it is easy – and it is revealing that it is so easy – to transpose her provocative analysis to the case of technological institutions.

The fact is that the regulation of technical change can be exploited and manipulated, as much as any other case of social regulation. 'The idea of citizen participation', says Arnstein without too many illusions, 'is a little like eating spinach; no one is against it in principle because it is good for you. Participation of the governed in their government is, in theory, the cornerstone of democracy – a revealed idea that is vigorously applauded by virtually everyone. The applause is reduced to polite handclaps, however, when this principle is advocated by the have-not blacks, Mexican-Americans, Puerto Ricans, Indians, Eskimos, and whites. And when the have-nots define participation as redistribution of power, the American consensus on the fundamental principles exploded into many shades or outright racial, ethnic, ideological, and political opposition.'

Let us leave aside these material have-nots and concentrate on the outcasts of knowledge – though the distinction is debatable since power depends as much on wealth as on knowledge. What Arnstein brings out, a point as true for the have-nots of the Welfare State as for those of what Don K. Price (1965) called 'the Scientific Estate', is that participation is hollow, frustrating and can lead to revolt when not accompanied by some redistribution of power. It merely allows the techno-structure to proclaim that all the 'interested parties' have been borne in mind whereas by the end of the process only a selected few have been at best consulted in order to safeguard the status quo. From the mere semblance of participation to the real thing, Arnstein distinguishes eight levels or rungs of a ladder, with each rung standing for a certain measure of citizen influence over the end-product of the decisions. By grouping them in two's she obtains three broad categories – non participation, tokenism and a genuine impact on the decision. (See Figure 13.1.)

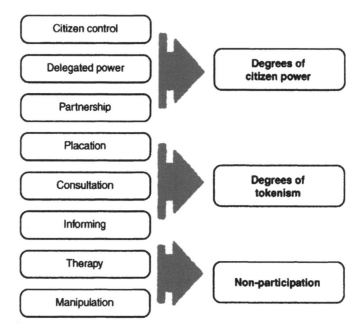

Figure 13.1 Arnstein's eight levels of participation
Source: S. R. Arnstein, 'A Ladder of Public Participation', *The Journal of American Institute of Planners*, July 1969, pp. 216–24.

The first category is manipulation and therapy. In the name of participation, an interest is taken in the concerns of ordinary people with the explicit purpose of 'educating' them and winning their support. This is really not more than a public relations stunt and can be illustrated by the hastily cobbled information campaigns after an industrial disaster (Bhopal) or near-disaster (Three Mile Island). The language of science is used to reassure the local population by imposing the therapy of experts. Just as, according to Arnstein, an effort is made to minister to the 'pathology' of the have-nots minorities rather than tackle the racism or inequalities underlying such a 'pathology', the arguments of these experts are focussed on the 'irrationality' of the fears occasioned by the big technologies rather than on the potential threats and real accidents which have engendered these threats.

At this level, the citizens are treated like children, handicapped, disabled or diseased persons condemned to beg for scientific assistance. Social control is reduced to a psychological insurance policy which attempts to tune attitudes and values to that 'hard facts' school of objectivity to which the techno structures belongs. In order to minister to such irrational fears and superstitious fantasies it turns to the social sciences and even psychoanalysis, which show how this opposition to new technologies is regressive, a rejection of the father, with elements of Oedipian repression and sexual inhibition. I am not joking. This was a common tone of most nuclear experts and institutions in the 1950s and later, as illustrated in a book published in French by a psychoanalyst, *The Atomic Anxiety and Nuclear Plants.*[2] In this book we are given extracts from the 1958 report of a WHO study group on 'the questions of mental health raised by the use of atomic energy for peaceful purposes which ends with this gem: 'From the viewpoint of Mental Health, the most satisfying solution for the future of peaceful applications would be the rise of a new generation which has learned to live with a certain degree of ignorance and uncertainty'.

Moving up from this masquerade or sham we come to the symbolic rite. Through informing, people are allowed to hear the arguments of the other side and to state their own views. They are no longer presumed to be 'mentally deranged' or children, but are treated as real ... adolescents. There is, of course, no guarantee that their fears and suggestions will be taken into consideration but the participation process has at least shed the guise of a socio-cultural insurance policy to take that of the classroom. At its best, although the information flows in a single direction, it is provided at an already advanced stage of the project. People are no doubt better informed, but the die has already been cast. In short, this kind of participation is like a formal lecture on the facts of the matter and the conclusions to be drawn in which the audience can do nothing but listen. It is certainly a step forward: better a pittance of information than none at all. Though tokenism may be a mere soporific, it shows the vices of democracy, so to speak, at least making some sort of gesture to its virtues.

At the next rung – consultation – people are invited to give their opinion. Aware of their rights and of their responsibilities, they intervene through opinion polls, attitude surveys or even in public inquiries. They do not, of course, intervene in the decision. As Arnstein rightly observes, they have 'participated in participation' and the techno-structure can now go ahead, happy with this formal contribution of the 'public' to its decision-making.

At a still higher level of tokenism, the rules of the game allow the have-nots to formulate their opinion. They are accorded a seat or two on the advisory committees and, provided that their occupants are accepted as recognized spokesman or women for the interests involved (e.g. have been elected by local associations), the opinions they voice may have a certain impact on the decision even though, for their part, the decision-makers insist on their full right to decide. This is placation. The have-nots feel less excluded and the techno-structure better understood.

From one rung of tokenism to the next, the progress is considerable and one cannot underestimate the benefits, not only for the public but for the policy-makers too. The insurance policy approach is repudiated, people are treated as adult citizens and above all, in this first attempt of real dialogue, the discussion makes room for fresh considerations, thus setting the issues in a more open, critical and pluralist context. At this level of participation, expert opinion no longer appears exclusively technical but most often than not turns out to be based on presuppositions which reveal the impossibility of drawing a sharp distinction between factual deductions and subjective, if not political, preferences. This is the first attempt to correct the imbalance between the decision-makers and those wanting to have a say in their decisions. By revealing that the arguments of the experts are not solely technical, the participation process abandons the esoteric, univocal and sacred land of 'scientific facts' for a more familiar terrain in which facts cannot be separated from values. The decision is taken in public rather than by the public, yet the social regulation of technical change becomes more than the semblance of a 'democratic dialogue' since the informing and learning process are no longer one way. At this level, there is discussion but still no negotiation.

One has to climb higher, to partnership, delegated power and especially citizen control, before the legitimacy of the decision is genuinely negotiated between the techno structure and the people. 'Partnership' implies that the groups contesting the decision have the financial resources to organize themselves and employ competent scientific, legal or other advisers and then see their views taken into account by the decision-makers. 'Delegated power' can be envisaged at the local or regional level when national interests are not at stake. Lord Ashby has cited one of the first cases of such a participation process. In 1975, the San Diego Gas and Electricity Company, having to build a thermal power-station in south California, invited the local people to set up a committee to examine possible sites, provided it with financial resources and technical advisers. This committee held its own public hearings and its recommendation was accepted.[3] Since then many other similar cases can be mentioned in Europe as well.

Thus lastly, the top rung of the ladder is 'citizen control', which may imply a referendum prepared long in advance by a public campaign, as was the case in Austria or Switzerland in relation to the building of nuclear plants or more recently genetic engineering in relation to transgenic plants and drugs.

In sum, if one wants really to go from the mere semblance of participation or from tokenism to real power-sharing, the amount of participation depends on the decision-makers accepting rules which are no longer laid down by themselves. In this context, what is at stake is simply the creditibility not only of the technical institutions concerned, but also of the whole democratic system.

'Credibility' is the keyword if I refer now to a report published by the US Department of Energy, which perfectly shows the link between transparency and the legitimacy of institutions. This report is indeed a vivid illustration of how reluctance to be open – or inability to practice transparency, i.e. to admit and recognize vulnerability – can lead a public institution to completely undermine its own creditibility with the public for the duration. *Earning Trust and Confidence: Requisites for Managing Radioactive Waste* is the title of this report which was commissioned by George Bush's Energy Secretary, Admiral James D. Watkins, and published in 1993 by his successor in the Clinton Administration, Mrs Hazel O'Leary. The fact that this report was asked for by a Republician and endorsed by a Democrat is in itself revealing. The title reflects the two aspects of the problem: how to ensure that the public puts its trust in the system and feels confident about it. This question, put by the Energy Secretary to the chairman of the Task Force, Todd La Porte, a professor of political science past Berkeley, touches on problems that are in many ways similar in scale to those in the former Soviet Union.[4]

In the United States, the Department of Energy is responsible for managing the nuclear wastes produced by both privately owned nuclear power stations generating electricity and by industries making nuclear weapons, including making good the damage to the environment at sites used for making and testing bombs. Managing these wastes absorbs more than a quarter of the Department's annual budget of over $2 billion, and the US has more than 3700 nuclear waste disposal sites, where the Department is only now starting to measure the levels of contamination. Some of these sites date back to the end of World War II and contain vast quantities of radioactive wastes from the postwar nuclear weapons programs. They have often been neglected for years, especially those at Hanford, Savannah River and West Valley.

Furthermore, the decision to create disposal sites deep underground where the waste would be left forever has naturally encountered vigorous opposition from the states, and even more from the countries, chosen from such sites, for instance Yucca Mountain. The decision has generated a controversy not only amongst scientists, but also amongst the various public agencies involved, with the Department of the Environment at odds with the Department of Energy. Opposition has also come from state governors, members of Congress and pressure groups, especially environmental organizations and Indian tribes whose lands were scheduled as possible sites.

That particular issue aside, the 'La Porte Report' also examined the whole history of the way the Department of Energy had behaved and it was extremely critical, saying that the Department was an excellent example of how a public institution could arouse suspicion, distrust and lack of credibility, not just on the part of environmental groups and local politicians in the states concerned, but even (to its great surprise) on the part of the industries involved in the programs of waste management. The Report notes that this distrust is not 'irrational' and cannot be dismissed as another example of NIMBY-ism (the 'not in my backyard' syndrome). Furthermore it will take a long time to reverse the trend. The Report offers a series of recommendations 'that are not just simply choices on a menu; (...) rather they represent the panel's recipe for what the Department should do to strengthen public trust and confidence'. These recommendations posit a package of conditions defined as necessary though still insufficient which, if implemented, might allow the Department of Energy eventually to become 'trustworthy' once more.

The Task Force was made up of social scientists, nuclear experts, industrialists, officials and representatives of voluntary organizations. The inquiry lasted two years, and during this time the Task Force sponsored two seminars, one organized by the National Academy of Sciences and the other by the National Academy of Public Administration, and it heard evidence from many people from both the public and private sectors. I mention this because it shows the Task Force's concern to present their report as a search for scientific legitimacy in a debate that must have given rise to considerable political tensions. Moreover, the principles stated as underlying the recommendations are presented as 'similar to geometric principles':

- 'Public trust and confidence is not a luxury. DoE not only has an obligation, but it also has a compelling need to do so.
- Public trust and confidence is not a one-way street. DoE must trust the public before it can expect the public to trust it. By the same token, the public and its

representatives must be held to a standard of behavior that is trustworthy.
- Under almost all circumstances currently relevant to DoE's waste management program, it is preferable to make decisions in an open, pluralistic forum than in a closed one that excludes actual or potential stakeholders.'

The recommendations run to several pages and call for changes not just in the relations between the Department of Energy and the public, but even more importantly in the Department's approach, methods and presentation with regard to managing nuclear waste – in short, a cultural revolution. The Task Force described 'the central contribution that trust and confidence makes to the legitimacy of public organizations within the American system of governance. That contribution derives from a democratic ideology that demands that public institutions operate in a transparent manner, that they adopt processes that not only permit but encourage broad segments of the public to participate, and that no segment finds itself permanently a loser in public controversies.'

The Report was a course criticized, like many another produced by the federal departments or any public administration, as being a way of achieving a clear conscience or making up for past errors. On the other hand, some people, including of course members of these departments, and in particular representatives of the nuclear lobby, were attacking its approach as unrealistic and doomed to be ineffectual. Let us put aside the question of how and how much time it will take to reduce such a gigantic amount of nuclear waste and just consider the implications of this lack of transparency. As the Task Force argues with respect to very simple, yet convincing, requirements of efficiency, 'in a very concrete sense, it is unlikely that agreements will be reached to manage radioactive wastes absent of a solid foundation of institutional trustworthiness'. I am ready to go further and say that there is a clear link between the suspicion that currently surrounds the functioning of democracies (and indeed the disenchantment with how they work felt by much of these electorate) and the suspicion of institutions whose activities are based on scientific rationality (Salomon, 1993).

Legitimacy crisis

In both cases, there is a crisis of legitimacy – in Habermas' sense – and each rebounds on the other. It is not a coincidence that Habermas' first work, his criticism of positivism, led him to question the reasoning behind the actions of the modern State: 'making politics more scientific' and 'making science more political' are both aspects of the same phenomenon whereby technology dominates everything else and is the source of new social conflicts and individual initiatives.[5] In all democracies the electorate's decreasing interest in politics is matched by the questioning or discrediting of institutions that are based on scientific rationality.

'Let no-one who is not a mathematician enter here' was, in Plato's world, an invitation to approach philosophy through science. In the contemporary technical system, a sign forbidding enter similar to the ones found on military areas warn off all those who do not know the password or the language of the tribe. The Goliaths of the big-technology networks, the public or private oligopolies of energy production, telecommunications, transport, space and especially defense-related industries are rarely vulnerable to the challenges from the Davids represented by pressure groups or individuals. As the controversy between Jürgen Habermas and Niklas Luhman shows, what is at stake is not just that the political system is becoming increasingly separated from modern society, it is how far individuals are able to make their voices heard in democracies.

For Luhman, transparency is a myth that is all the more dangerous as the gulf widens between open democracy and rationality. There is something old-fashioned in the 'old European view' that individuals can influence politics through their demands, when in fact the decision-making process is increasingly independent, administrative and technocratic. And the more that the electorate is urged to take a more active part, the more people are likely to feel frustrated; a decision-making process that pushed them to make explicit choices among the options would simply put them off trying at all. While Luhman highlights the growing independence of the machinery of government from civil society, Habermas is concerned about the increasing dependence of public institutions on powerful, well-organized economic interests.

They agree in emphasizing the limits imposed by the complexity of modern societies on the ability of citizens to make democracy more effective in practice, but whereas Luhman thinks that the aim of transparency is both useless and illusory, Habermas considers it to be an essential regulatory principle, 'an Idea of Reason' that democracies cannot abandon without ceasing to be democratic. Obviously this approach is the only one that avoids the double trap of technocracy and totalitarian rule. 'An Idea of Reason' is an ideal, and if one cannot achieve it, at least one should try to move in that direction: in any case, to merit their name democracies ought never to give up trying to attain that ideal.

Conclusion

This leads me to some concluding remarks about the experts' role in our societies. I mean to underline how little the process of expertise in relation to such technological issues can be dissociated from values judgments. To the techno-structure this may appear as undermining further the traditional image of science's neutrality. But this is the price to be paid by an effective democratic dialogue: the scientific advice appears ambiguous as soon as a controversy is extended to the public arena and reveals at the same time the ambivalence of scientists' loyalties. Scientists are trained not to trespass upon the world of judgment and political bickering, and yet they are called upon to take a stance on issues and decisions that cannot be restricted to a scientific assessment.

Take a current case of the threats to the earth's climate. Scientists may be alarmed personally at the possibility of catastrophic upheavals in climate change, but as experts they are divided and cannot come down publicly on the long-term trends toward warming because the evidence is not strong enough. Here we are no longer in the territory of science, but in a mixed field of scientific findings and political lobbying. If you read Ross Gelbspan's (1997) book, *The High Stakes Battle over Earth's Threatened Climate*, you see how much the oppositional view, though marginal, was able within the US Congress to challenge the work of the Intergovernmental Panel on Climate Change (IPCC) to the point that its conclusions appeared faulty and deliberately concocted. And within the fossil-fuel lobby in the US you find many scientists who may not seem worse from a scientific point of view than those who were members of the IPCC.

Indeed, the Summary of the IPCC report leaves out the reservations and presents the bald statements as an agreed conclusion. Moreover, it is striking to see how different the European approach is from the American one. For most of the European countries, this is now a settled matter and the precautionary principle should apply. But for the US, not only are there some 'greenhouse skeptics' in the scientific community, but even when most of the scientists agree on the reality of the threat, they doubt that the temperature rise in the near term can confidently be predicted and thus question the rationale for substantial and costly mitigation measures at this time. It is obvious that the development of policy on this issue in the US does not turn on the scientific evidence, but on political and economic forces which challenge the Congress, the Administration and the President's and Vice President's views.[6]

Uncertainty on one side, possible irreversibility on the other. This is an issue where the interests of the industrial sector play a role in the policy process as much as political and public opinion in relation to the environment. The latter pushes governments to a strong stance in favor of fuel taxes. The former tends to rely on market mechanisms for achieving cost-effective emissions. After the Rio de Janeiro, Kyoto and Buenos Aires United Nations conferences, we know that the matter is far from being settled from a political point of view and that the scientific community is still divided today.

Whatever the results of future negotiations, this is a perfect example of the ambiguous role that scientists are today doomed to play. And this is really new if one thinks of the predicament of the Charter of the Royal Society, which was given the objective of, on the one hand, 'perfecting the knowledge of natural things', and on the other, 'not meddling with Metaphysics, Morall and Politics' as Sandy Lakoff so beautifully opened our discussions here. In our days, scientists cannot avoid being involved as experts in scientific matters that are exposed to the pressure of interests and passions characteristic of the political scene and therefore they are directly immersed in the controversial area of 'politics and morals'.

Such dispute between the pros and cons is no longer a scientific controversy as such, nor should the experts' stand in such controversies be defined as presenting 'pure' scientific statements. When experts are brought in, their role does not consist in acting as arbiters, but rather they feature in stressing the facts, the consequences and the issues which are at stake and which highlight the sources of controversy and conflict. This does not mean that they themselves keep out of the fray. It is one thing to produce scientific knowledge. It is quite another to use it for the purposes of decision-making. As Philippe Roqueplo has shown clearly in connection with the problems related to the greenhouse effect and indeed to environmental problems in general, the credibility of both experts and science is compromised by the different procedures.[7]

What transforms a scientific statement into an expert opinion is its use in the decision-making process. From then on the scientist cannot avoid bringing something of himself or herself into the process: his or her personal and professional background, his or her sympathies and values, and so on. The objectivity expected of him or her has to be maintained not only despite pressures from the public or from government but also, no matter how honest he or she is, despite his or her own prejudices and commitment. *In short, a clear distinction must be made between three separate functions: producing knowledge, offering an expert opinion and making decisions.* The scientists' role is to advance the state of knowledge, though with the recognition that knowledge is always under scrutiny as is being distributed across public. The experts' role is to highlight the issues by providing the relevant information, and to be prepared to enter the realm of controversy. And the policy-makers' role is to weigh up the information and to make decisions, accepting the sanction of public scrutiny and the possibility of being removed from office.

It is important not to muddle the different roles the various areas of competence, or – even more – the areas of responsibility. The three elements involve different but overlapping areas of debate, and all three are essential inputs in the process of public decision-making, which is not really a matter of scientific objectivity, but of the clash between strong feelings and interests. In brief, one needs to channel the latent violence by giving it a chance to be transformed into an expression of public will. For this to occur, the decisions must at least be open to negotiation and, for a start, the relevant information must be made available from the very beginning of a project, and not when it is too late for the plans to be changed. But another condition is that the pressure groups have enough money to organize their campaigns and hire competent advisers (scientists, lawyers, etc.). In short, they should not be at a disadvantage compared with the businesses, public or private, whose decisions they oppose, and this may mean that the businesses concerned will have to contribute to funding counter-investigations.

To conclude: it is never enough for individuals to have the right simply to oversee the actions of a government department or to have access to information held by that department. More is required for real transparency. Something has to be done to correct the inevitable imbalance between the strength of those taking the decisions and the concerns of the individuals or lobbies trying to influence those decisions. In short, it is a matter of renegotiating and redistributing power. Transparency rises above the easy jibes directed against it when it is clearly understood that the regulation of technologies cannot be left entirely to technicians. Experts should not have the final say in a debate where the consequences are never purely technical. This view, I think, permeates every fibre of this book.

Notes

1. M. Foucault, quoted in J. Theys, 'Postface', Conquête de la sécurité, gestion des risques, C. Douriens, J.-P., Gaillard *et al.* (eds), L'Harmattan, Paris, 1991.
2. Mendel, G. and Guedeney, C. (1978) *L'angoisse atomique et les centrales nucléaires.* Paris: Payot.
3. Lord Ashby, *Nature*, in his review of Guild Nichol's report on technology and public participation published by OECD in 1979, Technology on Trial, no. 283, February 1980.
4. T. La Porte, Earning, Trust and Confidence: Requisites for Managing Radioactive Waste, Final Report of the Secretary of Energy's Advisory Board, Task Force on Radioactive Waste Management, US GPO, Washington, 1993.
5. Habermas, J. (1975) *Legitimation Crisis.* Boston: Beacon Press; and his dialogue with N.

Luhman, Theorie des Gesellschaft oder Sozialtechnologie?, Surkamp, Franckfurt, 1971.

6. Skolnikoff, E. B. (1997) Same Science, Differing Policies: The Saga of Global Climate Change. MIT Joint Program on Science and Policy of Global Change, Report no. 22, August 1997; and From Science to Policy: The Science-Related Politics of Climate Change Policy in the U.S., Report no. 46, January 1999.

7. Roqueplo, P. (1993) *Climats sous surveillance – Limites et conditions de l'expertise scientifique. Economica.* See also the review of this book by A.-J. Guérin in the journal of the Ecole polytechnique, La jaune et la rouge, November 1993.

References

Abramovitz, M. (1989) *Thinking About Growth*. Cambridge: University Press.

Abramovitz, M. (1994) The origins of the postwar catch-up and convergence boom. In J. Fagerberg, N. von Tunzelman and B. Verspagen (eds), *The Dynamics of Technology, Trade and Growth*. London: Edward Elgar.

Abramson, J., Arterton, F. and Orren, G. (1988) *The Electronic Commonwealth*. New York: Basic Books.

Alexander, A. J., Hill, P. T. and Bodilly, S. J. (1989) The Defense Department's Support of Industry's Independent Research and Development (IR&D), R-3649-ACQ. Santa Monica, CA: RAND.

Alic, J., Branscomb, L. M., Brooks, H., Carter, A. B. and Epstein, G. (1992) *Beyond Spinoff*. Boston: Harvard Business School Press.

Altmann, J., Liebert, W., Neuneck, G. and Schreffan, J. (forthcoming) Dual-use and the conversion of military-related R&D in Germany. In J. Rotblat, J. Holdren, J. Reppy and V. Avduyevsky (eds), *Converting Military R&D*. London: Macmillan.

Alperovitz, G. (1965) *Atomic Diplomacy*. New York: Simon and Schuster.

Alsop, Joseph and Stewart (1954) *We Accuse! The Story of the Miscarriage of Justice in the Case of J. Robert Oppenheimer*. New York: Simon and Schuster.

Amable, B. (1992) Effects d'apprentissage, compétitivité, hors-prix et croissance cumulative. *Economie Appliquée*, **45**.

Amable, B. (1993) National effects of learning international specialization and growth paths. In D. Foray and C. Freeman (eds), *Technology and the Wealth of Nations*. London: Pinter.

Amsden, A. (1989) *Asia's Next Giant: South Korea and Late Industrialization*. New York: Oxford University Press.

Anderson, A. M. (1984) *Science and Technology in Japan*. Harlow, Essex: Longman.

Anderson, R., Bikson, T., Law, S. and Mitchell, B. (1995) *Universal Access to Email: Feasibility and Social Implications*. Santa Monica, CA: Rand Corporation.

Angell, M. (1996) *Science on Trial*. New York: Norton.

Archer, C. (1983) *International Organizations*. George Allen & Unwin.

Archibugi, D. and Pianta, M. (1992) *The Technological Specialization of Advanced Countries*. Dordrecht: Kluwer.

Arnstein, S. R. (1969) A ladder of public participation. *The Journal of American Institute of Planners*, July.

Aron, R. (1954) *The Century of Total War*. Garden City, NY: Doubleday.

Arterton, F. (1987) *Teledemocracy: Can Technology Protect Democracy?* Beverly Hills, CA: Sage.

Augustine, N. R. (1986) *Augustine's Laws*. New York: Penguin Books.

Australian Science and Technology Council (1995) *Surf's Up: Alternative Futures for Full Service Networks in Australia*. Canberra: AGPS.

Badash, L. (1985) *Kapitza, Rutherford and the Kremlin*. New Haven, CT: Yale University Press.

Bagli, C. V. (1997) Two giants join a merger parade in arms industry, *New York Times*, 4 July.

Balmer, B. (1996) Managing Mapping in the Human Genome Project. *Social Studies of Science*, **26**.

Balogh, B. (1991) *Chain Reaction: Expert Debate and Public Participation in American Commercial Nuclear Power, 1945–1975*. New York: Cambridge University Press.

Baran, P. (1956) *The Political Economy of Growth*. New York: Monthly Review Press.

Barber, B. (1962) *Science and the Social Order*. New York: Collier.

Barber, B. J. (1984) *Strong Democracy: Participatory Politics for a New Age*. Berkeley: University of California Press.

Bardhan, P. (1996) The political economy of development policy: an Asian perspective. Development Thinking and Practice Conference, Washington, DC, 3–5 September 1996.

Baumgartner, F. R. and Wilsford, D. (1994) Science within the state. In Etel Solingen (ed.) *Scientists and the State: Domestic Structures and International Context*. Ann Arbor: University of Michigan Press.

Barzelay, M. (1992) *Breaking Through Bureacracy*. Berkeley: University of California Press.

Beard, E. (1976) *Developing the ICBM: A Study in Bureaucratic Politics*. New York: Columbia University Press.

Benedick, R. E. (1991) *Ozone Diplomacy: New Directions in Safeguarding the Planet*. Cambridge, MA: Harvard University Press.

Bernal, J. D. (1939) *The Social Function of Science*. London: Routledge.

Berry, M. J. (ed.) (1988) *Science and Technology in the USSR*. Harlow, Essex: Longman.

Bessant, J. and Dodgson, M. (1996) *Effective Innovation Policy: A New Approach*. New York: Routledge.

Beyerchen, A. (1977) *Scientists Under Hitler*. New Haven, CT: Yale University Press.

Bimber, B. (1996) *The Politics of Expertise in Congress: The Rise and Fall of the Office of Technology Assessment*. Albany: State University of New York Press.

Bischak, G. A. (1989) *The Implications of Alternative Security Doctrines*. London: Routledge, p. 31.

Boffey, P. M. (1975) *The Brain Ban of America: An Inquiry into the Politics of Science*. New York: McGraw Hill.

Boggio, L. (1993) Growth and international competitiveness in a Kaldorian perspective. Istituto di Teoria Economica e Metodi Quantitativi, Università cattolica del Sacro Cuore, mimeo.

Bonchek. M. (1995) Grassroots in Cyberspace: Using Computer Networks to Facilitate Political Participation. The Political Participation Project, MIT, working paper, 95–2.2.

Bortnick, J. and Miller, N. R. (1986) Scientific and technical information. In Albert H. Teich and Jill H. Pace (eds), *Science and Technology in the USA*. London: Longman.

Branscomb, L. M., Kodama, F. and Florida, R. (eds) (1999) *Industrializing Knowledge*. Cambridge, MA: MIT Press.

Brinkley, J. (1997) *Defining Vision: The Battle for the Future of Television*. New York: Harcourt Brace.

Broadband Services Expert Group (1994) *Networking Australia's Future*. Canberra: AGPS.

Brodie, B. (1946) *The Absolute Weapon: Atomic Power and World Order*. New York: Harcourt Brace.

Brodkin, E. (1992) The organization of disputes: the bureaucratic construction of welfare rights and wrongs. *Studies in Law, Politics, and Society*, **12** (A).

Brooks, J. (1964) The scientific advisor. In R. Gilpin and C. Wright (eds) *Scientists and National Policymaking*. New York: Columbia University Press.

Brooks, H. (1968) *The Government of Science*. Cambridge, MA: MIT Press.

Brooks, H. (1994) The relationship between science and technology. *Research Policy*, **23**.

Brown, J. S. (1995) Contribution in report on: Workshop to Assess Corporate Research Restructuring. National Academy of Sciences.

Brown, L. (1970) *Seeds of Change: The Green Revolution and Development in the 1970s*. New York: Praeger.

Brzoska, M., Wilke, P. and Wulf, H. (1996) The changing civil-military production mix in Western Europe's defense industry. Paper prepared for the Council on Foreign Relations Study Group on International Downsizing, Arms Exports and Conversion, New York, October 1996 (mimeo).

Burnham, D. (1983) *The Rise of the Computer State*. New York: Random House.

Bush, V. (1945) *Science, the Endless Frontier*. Washington DC: Public Affairs Press.

Cabinet Office (1995) *Report of the Technology Foresight Panel on Defence and Aerospace*. London: HMSO.

Cahn, A. (1971) *Eggheads and Warheads: Scientists and the ABM*. Cambridge, MA: Science and Public Policy Program.

Callon, S. (1995) *Divided Sun: MITI and the Breakdown of Japanese High-Tech Industrial Policy, 1975–1993*. Stanford: Stanford University Press.

Cambrosio, A., Limoges, C. and Hoffman, E. (1992) Expertise as a network: a case study of the controversies over environmental release of genetically engineered organisms. In N. Stehr and R. V. Ericson (eds), *The Culture and Power of Knowledge: Inquiries into contemporary societies*. New York and Berlin: Walter de Gruyter.

Canter, U. and Hanusch, H. (1990) Process and product innovation in an international trade context. Paper presented at the EAEPE-Conference, Florence, Economics of Innovation and New Technology.

Cantwell, A. J. (1991) The theory of technological competence and its application to international production. In D. G. McFetridge (ed.), *Foreign Investment, Technology and Growth*. Calgary, Canada: University of Calgary Press.

Cantwell, J. (1989) *Technological Innovation and Multinational Corporations*. Oxford: Basil Blackwell.

Canuto, O. (1994) *Brasil e Coréia do Sul: os (des) caminhos da industrialização tardia*. São Paulo: Nobel.

Carson, R. (1962) *Silent Spring*. Boston, MA: Houghton Mifflin.

Cimoli, M. (ed.) (2000) *Developing Innovation Systems: Mexico in the Global Context*. London: Continuum.

Cimoli, M. and Soete, L. (1992) A generalized technological gap trade model. *Economie Appliquée*, **45**.

Cimoli, M., Dosi, G. and Soete, L. (1993) Innovation diffusion institutional differences and patterns of trade: a North-South model. In P. A. David, G. Dosi and F. Arcangeli (eds), *Technology Diffusion and Economic Growth. International and National Policy Perspectives*, **3**, Oxford: Oxford University Press. (DRC Paper no. 36, SPRU, University of Sussex).

Clark, R. W. (1965) *Tizard*. Cambridge, MA: MIT Press.

Clegg, S. (1989) *Frameworks of Power*. London: Sage.

Cochrane, R. (1976) *Measures for Progress: A History of the National Bureau of Standards*. Arno.

Coghlan, A. (1994) Stony ground for Britain's ploughshares. *New Scientist*, 22, January.

Cohen, L. R. and Noll, R. G. (1991) *The Technology Pork Barrel*. Washington: Brookings.

Cole, L. (1983) *Politics and the Restraint of Science*. Totowa, NJ: Rowman and Allenheld.

Commoner, B. (1971) *The Closing Circle: Nature, Man, and Technology*. New York: Alfred Knopf.

Cooper, P. (1996) DOD repackages dual-use goals. *Defense News*, **11**, 15 April.

Cozzens, S. E. (1990) Autonomy and power in science. In S. E. Cozzens and T. F. Gieryn (eds), *Theories of Science in Society*. Bloomington: Indiana University Press.

Cozzens, S. E. and Melkers, J. (1998) Use and usefulness of performance measurement in state science and technology programs. *Policy Studies Journal*.

Crewdson, J. (1993) Perky cheerleaders. *Nieman Reports*, Winter.

Cronberg, T. Concepts of military technology: contesting the boundaries between the civilian and the military. In P. Gummett, M. Boutoussov, J. Farkas, and A. Rip (eds), *Military R&D after the Cold War: Conversion and Technology Transfer in Eastern and Western Europe*. Dordrecht: Kluwer Academic Publishers.

Crow, M. (1994) *Design and Analysis*. Washington: NAE.

Cunningham, P. (ed.) (1999) *Science and Technology in the United Kingdom*. London: Cartermill.

Darvaes, G. (ed.) (1988) *Science and Technology in Eastern Europe*. London: Longman.

Dasgupta, P. and David, P. (1994) Toward a new economics of science. *Research Policy*, **23**.

de Briganti, G. (1996) French reform calls for weapon guarantees. *Defense News*, September 16–22.

de la Mothe, J. (1992) *C. P. Snow and the Struggle of Modernity*. Austin: University of Texas Press.

de la Mothe, J. and Ducharme, L. M. (1990) *Science, Technology and Free Trade*. London: Pinter.

de la Mothe, J. and Dufour, P. (1993) *Science and Technology in Canada*. London: Longman.

de la Mothe, J. and Dufour, P. (1995) Is science policy in the doldrums? *Nature*, **374**.

de la Mothe, J. and Paquet, G. (1995) 'Circumstantial evidence': a note on Canadian science policy. *Science and Public Policy*, **22**, (6).

de la Mothe, J. and Paquet, G. (1996) The technology–trade nexus. *Technology in Society*, **18**, (2).

de la Mothe, J. and Paquet, G. (eds) (1996) *Evolutionary Economics and the New International Political Economy*. London: Pinter.

de la Mothe, J. and Paquet, G. (eds) (1998a) *Local and Regional Systems of Innovation*. Boston, MA: Kluwer.

de la Mothe, J. and Paquet, G. (1998b) Structural competitiveness and interdependencies: regional patterns. In J. H. Dunning and G. Boyd (eds) *Structural Change and Cooperation in the Global Economy*. Cheltenham: Edward Elgar.

de la Mothe, J. and Paquet, G. (eds) (1999) *Information, Innovation and Impacts*. Boston, MA: Kluwer.

de Sola Pool, I. (1983) *Technologies of Freedom*. Cambridge, MA: Belknap Press.

de Vestel, P. (1997) 'Belgium'. In P. Gummett and J. Stein (eds) *European Defence Technology in Transition*. Amsterdam: Harwood.

DeBeer, G. (1960) *The Sciences Were Never at War*. London: Thomas Nelson and Sons.

Della Giusta, M. (1996) Education and Technological Capabilities in Industrialization. Master's Thesis, University of Reading.

Department of Industry Technology and Tourism (DIST) (1996) *Australian Business Innovation: A Strategic Analysis*. Canberra: AGPS.

Dickson, D. (1997) Labour's science appointment lifts hope for reform. *Nature*, May 8.

Doern, G. B. (1972) *Science and Politics in Canada*. Montreal: McGill and Queen's University Press.

Dooley, J. J. (1997) Unintended consequences: energy R&D in a deregulated market. PNNLSA-28561, Pacific Northwest National Laboratory, Washington, DC, February. Holdren, *op. cit.*

Dosi, G. and Soete, L. (1983) Technological gaps and cost-based adjustment: some explorations on the determinants of international competitiveness. *Metroeconomica*, **12**.

Dosi, G. and Freeman, C. (1992) The diversity of development patterns: on the processes of catching-up, forging ahead and falling behind. Paper presented at Conference on Economic Growth and the Structure of Long-term Development, Varenna 1–3 October.

Dosi, G., Pavitt, K. and Soete, L. (1990) *The Economics of Technical Change and International Trade*. London and New York: Harvester Wheatsheaf Press, and New York University Press.

Dupré, J. S. and Lakoff, S. A. (1963) *Science and the Nation: Policy and Politics*. Englewood Cliffs, NJ: Prentice Hall.

Dupree, A. H. (1987) *Science in the Federal Government*. Baltimore: Johns Hopkins University Press.

Easterbrook, G. (1995) *A Moment on the Earth: The Coming Age of Environmental Optimism*. Harmondsworth, Middlesex: Penguin Books.

England, J. M. (1983) *A Patron for Pure Science: The National Science Foundation's Formative Years, 1945–57*. Washington, DC: National Science Foundation.

Epstein, S. (1996) *Impure Science: AIDS, Activism, and the Politics of Knowledge*. Berkeley, CA: University of California Press.

Ergas, H. (1987) Does technology policy matter? In B. R. Guile and H. Brooks (eds), *Technology and Global Industry*. Washington: National Academy Press.

Evenson R. E. (1982) Agriculture. In R. R. Nelson (ed.) *Government and Technical Progress: A Cross-Industry Analysis*. New York: Pergamon Press.

Ezrahi, Y. (1990) *The Descent of Icarus: Science and the Transformation of Contemporary Democracy*. Cambridge, MA: Harvard University Press.

Fagerberg, J. (1995) Globalization and convergence. *Technological Forecasting and Social Change*, **47**, (2).

Faulkner, W. and Senker, J. (1994) Making Sense of Diversity: Public-private sector research linkages in three technologies. *Research Policy*, **23**.

Faulkner, W. and Senker, J. (1995) *Knowledge Frontiers: Public Sector Research and Industrial Innovation in Biotechnology, Engineering Ceramics and Parallel Computing*. New York: Oxford University Press.

Feigenbaum, H., Weaver, R. K. and Samuels, R. (1997) Innovation, coordination, and implementation in energy policy. In R. K. Weaver and B. A. Rockman (eds) *Do Institutions Matter?* Washington: Brookings.

Ferné, G. (1989) *Science and Technology in Scandinavia*. Harlow, Essex: Longman.

Foray, D. and Gibbons, M. (1996) Discovery in the context of application. *Technological Forecasting and Social Change*, **53**, 3.

Forje, J. W. (1989) *Science and Technology in Africa*. Harlow, Essex: Longman.

Freedman, L. (1981) *The Evolution of Nuclear Strategy*. New York: St. Martin's Press.

Freedman, R. (1999) A review of the networks of Centres of Excellence. Toronto: The Impact Group, mimeo.

Freeman, C. (1963) The Plastic Industry: a comparative study of research and innovation. *National Institute Economic Review*, **26**.

Freeman, C. (1987) *Technology Policy and Economic Performance: Lessons from Japan*. London: Francis Pinter.

Freeman, C. (1988) Japan: a new national system of innovation? In G. Dosi *et al.* (eds) *Economic Theory and Technical Change*. London: Pinter.

Freeman, C. (ed.) (1990) *The Economics of Innovation*. Brookfield, VT: E. Elgin.

Friedman, T. L. (1999) *The Lexus and the Olive Tree*. New York: Farrar, Straus and Giroux.

Fukuyama, F. (1999) *The Great Disruption: Human Nature and the Reconstruction of Social Order*. New York: Free Press.

Galbraith, J. K. (1967) *The New Industrial State*. New York: Mentor Books.

Ganley, G. D. (1991) Power to the people via personal electronic media. *Washington Quarterly*, 5–22.

Gelbspan, R. (1997) *The High Stakes Battle over Earth's Threatened Climate*. Addison-Wesley.

Gibbons, M., Limoges, C., Nowotny, H., Schwarzman, S., Scot, P. and Trow, M. (1994) *The New Production of Knowledge: the Dynamics of Science and Research in Contemporary Science*. London: Sage Publications.

Gilpin, R. (1968) *American Scientists and Nuclear Weapons Policies*. Princeton: Princeton University Press.

Godwin, R. (1988) *One Billion Dollars of Influence: The Direct Marketing of Politics*. New Jersey: Chatham House.

Golden, W. T. (ed.) (1991) *Worldwide Science and Technology Advice*. New York: Pergamon Press.

Gomulka, S. (1971) Inventive activity, diffusion and the stages of economic growth, Aarthus, Skrifter fra Aarthus Universtets Okonomisske Institut 24.

Gore, A. (1997) The metaphor of distributed intelligence. *Science*, **272**, 177.

Graham, L. (1967) *The Soviet Academy of Sciences and the Communist Party, 1927–1932*. Princeton: Princeton University Press.

Granstrand, O., Hakanson, L. and Sjolander, S. (1993) Internationalization of R&D – a survey of some recent research. *Research Policy*, **22**.

Greenberg, D. S. (1967) *The Politics of Pure Science*. New York: New American Library.

Greenleaf, G. (1991) Can the data matching epidemic be controlled? *The Australian Law Journal*, **65**, 4.

Greenwood, M. R. C. (1996) Desperately seeking friends. *Science*, **272**, 933.

Griffin, K. and Gurley, J. (1985) Radical analyses of imperialism: the Third World and the transition to socialism. *Journal of Economic Literature*, **23**.

Grose, L. and Levitt, B. (1994) *The Higher Superstition*. Washington, DC: John Hopkins University Press.

Gummett, P. and Reppy, J. (eds) (1988) *The Relations Between Defence and Civil Technologies*. Dordrecht: Kluwer.

Gummett, P. and Reppy, R. (1990) Military industrial networks and technical change. *Government and Opposition*, **25**.

Gummett, P. and Stein, J. (eds) (1997) *European Defence Technology in Transition*. Amsterdam: Harwood.

Guston, D. (1996) Principal-agent theory and the structure of science policy. *Science and Public Policy*, **23**.

Haas, E. B. (1964) *Beyond the Nation-State: Functionalism and International Organizations*. Palo Alto: Stanford University Press.

Haas, P. M., Keohane, R. O. and Levy, M. A. (1993) *Institutions for the Earth: Sources of Effective International Environmental Protection*. Cambridge, MA: MIT Press.

Haberer, J. (1969) *Politics and the Community of Science*. New York: Van Nostrand Reinhold.

Hackman, S. (1997) And now a word from your neighbors. *Technology Review*, **6**.

Handy, C. (1994) *The Age of Paradox*. Cambridge, MA: Harvard Business School Press.

Hart, D. M. (1997) Making technology policy in the White House. In L. M. Branscomb (ed.) *Investing in Innovation*. Cambridge, MA: MIT Press.

Hart, D. (forthcoming) *Forging the 'Postwar Consensus': Science, Technology, and Economic Policy in the U.S., 1921–1953*. Princeton, NJ: Princeton University Press.

Hart, D. and Victor, D. (1993) Scientific elites and the making of US policy for climate change research, 1957–74. *Social Studies of Science*, **23**.

Hawkins, H. S., Greb, G. A. and Weiss-Szilard, G. (eds) (1987) *Toward a Livable World: Leo Szilard and the Crusade for Nuclear Arms Control*. Cambridge, MA: MIT Press.

Hays, S. P. (1987) *Beauty, Health, and Permanence: Environmental Politics in the United States, 1955–1985*. New York: Cambridge University Press.

Heisenberg, W. (1971) *Physics and Beyond: Encounters and Conversations*, trans. A. Pomerans.

New York: Harper and Row.

Herken, G. (1985) *Counsels of War*. New York: Alfred Knopf.

Herken, G. (1992) *Cardinal Choices: Presidential Science Advising from the Atomic Bomb to SDI*. New York: Oxford.

Hillman, D. J. (1982) Decision making with modern information and communications technology: opportunities and constraints. In Albert H. Teich and Ray Thornton (eds), *Science, Technology, and the Issues of the Eighties: Policy Outlook*. Boulder, CO: Westview.

Hirsch, S. (1965) The US electronics industry in international trade. *National Institute Economic Review*, **34**.

Hirschman, A. O. (1971) Ideologies of economic development in Latin America. In A. O. Hirschman (ed.) *A Bias for Hope*. New Haven, CT: Yale University Press.

Hirschman, A. O. (1977) Generalized linkage approach to development, with special reference to staples. In M. Nash (ed.) *Essays on economic development and cultural change in honour of B. F. Hoselitz*. Chicago: University of Chicago Press.

HMSO (1972) *The Organisation and Management of Government Research*. London: The Rothschild Report, Cmnd. 4814.

Hobday, M. (1995) *Innovation in East Asia*. Cheltenham: Edward Elgar.

Holdren, J. P. (1992) Radioactive waste management in the United States: evolving policy prospects and dilemmas. *Annual Review of Energy and Environment*, **17**.

Holdren, J. P. (1997) Federal energy research and development for the challenges of the 21st century. In L. M. Branscomb (ed.) *Investing in Innovation*. Cambridge, MA: MIT Press.

Holloway, D. (1994) *Stalin and the Bomb: The Soviet Union and Atomic Energy 1939–1956*. New Haven, CT: Yale University Press.

Hood, C., Roberts, P. and Chilvers, M. (1990) Cutbacks and public bureaucracy: consequences in Australia. *Journal of Public Policy*, **10**, 2.

Hornig, D. F. (1968) Address at Carnegie-Mellon University. *Science*, **161**.

Houghton, J., Pucar, M. and Knox, C. (1996) *Mapping the Information Industries*, Staff Information Paper, Productivity Commission.

Huber, P. (1994) *Orwell's Revenge: The 1984 Palimpset*. New York: Free Press.

Huber, P. and Kellogg, M. (1994) *The Geodesic Network II: 1993 Report on Competition in the Telephone Industry*. Washington, DC: The Geodesic Company.

Hufbauer, G. (1966) *Synthetic Materials and the Theory of International Trade*. London: Buckworth.

Huff, C., Sproull, L. and Kiesler, S. (1989) Computer communication and organisational commitment: tracing the relationship in a City government. *Journal of Applied Psychology*, **19**, 16.

Huxley, A. (1932) *Brave New World*. New York: Harper and Row.

Intergovernmental Panel on Climate Change (1995) Summary for Policymakers. http: // www.unep.ch/ipcc

Intersociety Working Group (1996) *Research and Development: AAAS Report XXI, FY 1997*. Washington: American Association for the Advancement of Science.

Irving, D. (1967) *The German Atomic Bomb: The History of Nuclear Research in Nazi Germany*. New York: Simon and Schuster.

Jacobson, H. K. (1984) *Networks of Interdependence: International Organizations and the Global Political System*, 2nd edn. New York: Alfred A. Knopf.

Jacobson, H. K. and Stein, E. (1966) *Diplomats, Scientists and Politicians*. Ann Arbor: Michigan University Press.

Jasanoff, S. (1990) *The Fifth Branch: Science Advisers as Policy Makers*. Cambridge, MA: Harvard University Press.

Jasanoff, S. and Ilgen, T. (1985) *Controlling Chemicals: The Politics of Regulation in Europe and the United States*. Ithaca, NY: Cornell University Press.

Johnson, C. (1982) *MITI and the Japanese Miracle*. Stanford: Stanford University Press.

Jones, G. (1971) *The Role of Science and Technology in Developing Countries*. London: Oxford University Press.

Joravsky, D. (1961a) *Soviet Marxism and Natural Science, 1917–1932*. New York: Columbia University Press.

Joravsky, D. (1961b) *The Lysenko Affair*. Cambridge, MA: Harvard University Press.

Joseph, R. (1993) The politics of telecommunications reform: a comparative study of Australia and New Zealand. University of Wollongong, Science and Technology Analysis Research Program, Working Paper No. 12.

Josephson, P. R. (1991) *Physics and Politics in Revolutionary Russia*. Berkeley: University of California Press.

Justman, M. and Teubal, M. (1996) Technological infrastructure policy (TIP): creating capabilities and building markets. In M. Teubal, D. Foray, M. Justman and E. Zuscovitch (eds), *Technological Infrastructure Policy: An International Perspective*. The Jerusalem Institute for Israel Studies. Dordrecht: Kluwer.

Kahin, B. (1997) Beyond the national information infrastructure: technology-informed policy and policy-enabling technology. In L. M. Branscomb (ed.) *Investing in Innovation*. Cambridge, MA: MIT Press.

Kantrowitz, A. (1967) Proposal for an Institution of Scientific Judgement. *Science*, **156**.

Kaplan, F. M. (1983) *The Wizards of Armageddon*. New York: Simon and Schuster.

Katz, J. (2000), Structural reforms, the sources and nature of technical change and the functioning of the national systems of innovation: the case of Latin America. In D. Nelson and L. Kim (eds), *Systems of Innovation*. Oxford: Oxford University Press. Paper presented at the International Symposium on Innovation and Competitiveness in NIEs, Seoul, Korea, May 1997.

Keller, S. (1963) *Beyond the Ruling Class: Strategic Elites in Modern Society*. New York: Random House.

Kellerman, E. W. (1988) *Science and Technology in France and Belgium*. Harlow, Essex: Longman.

Kelley, M. and Watkins, T. (1995) In from the cold: prospects for conversion of the defense industrial base. *Science*, **268**, 28 April.

Kennedy, P. (1987) *The Rise and Fall of the Great Powers: Economic Change and Military Conflict From 1500 to 2000*. New York: Random House.

Kenny, M. (1996) The role of information, knowledge and value in the late 20th century. *Futures*, **28**, 8.

Kevles, D. (1977) *The Physicists: The History of a Scientific Community in Modern America*. New York: Alfred Knopf.

Keynan, A. (1991) The U.S. As a Partner in Scientific and Technological Cooperation: Some Perspective From Across the Atlantic. Carnegie Commission on Science, Technology, and Government, June.

Kickert, W. (1996) Expansion and diversification of public administration in the postwar welfare state: the case of the Netherlands. *Public Administration Review*, **56**, 1.

Kim, L. (1997a) Korea's national innovation system in transition. Paper presented at the International Symposium on Innovation and Competitiveness in NIEs, Seoul, Korea, May 1997.

Kim, L. (1997b) *Imitation to Innovation: The Dynamics of Korea's Technological Learning*. Cambridge, MA: Harvard Business School Press.

King, J. and Kraemer, K. (1986) Computing and public organisations. *Public Administration Review*, **46**.

Kirkpatrick Sale (1996) *Rebels Against the Future: Luddites and Their War on the Industrial Revolution: Lessons for the Computer Age*. New York: Perseus Press.

Klaw, S. (1968). *The New Brahmins: Scientific Life in America*. New York: William Morrow.

Kodama, F. (1991) *Analyzing Japanese Innovation*. London: Pinter.

Kourislky, P. (1998) *La science en partage*. Odile Jacob: Paris.

Kraemer, K., King, J., Dunkle, D. and Lane, J. (1989) *Managing Information Systems: Change and Control in Organisational Computing*. San Francisco: Jossey Bass.

Krementsov, N. (1997) *Stalinist Science*. Princeton: Princeton University Press.

Krugman, P. (1987) The narrow moving band, the Dutch disease and the competitive consequences of the Mrs Thatcher. *Journal of Development Economics*, **27**.

Krugman, P. (1999) *The Return of Depression Economics*. London: Allen Lane.

Ladd, E. C. and Lipset, S. M. (1972) Politics of academic scientists and engineers. *Science*, **76**, 9.

Lakoff, S. (1966) The third culture: science in social thought. In S. Lakoff (ed.) *Knowledge and Power: Essays on Science and Government*. New York: Free Press.

Lakoff, S. (1975) Scientists, technologists and political power. In Ina Spiegel-Rosing and Derek de Sola Price (eds) *Science, Technology and Society: A Cross-Disciplinary Perspective*. Beverly Hills, CA: Sage Publications.

Lakoff, S. (ed.) (1979) *Scientists and Ethical Responsibility*. Reading, MA: Addison-Wesley.

Lakoff, S. and York, H. F. (1989) *A Shield in Space? Technology, Politics and the Strategic Defense Initiative*. Berkeley: University of California Press.

Lall, S. (1997) Technological change and industrialisation in the Asian NIEs: achievements and challenges. Paper presented at the International Symposium on Innovation and Competitiveness in NIEs, Seoul, Korea, May 1997.

Lane, N. (1998) Remarks to the AAAS Science and Technology Policy Colloquium, 1 May.

Lapp, R (1965) *The New Priesthood: The Scientific Elite and the Uses of Power*. New York: Harper and Row.

Latin American Newsletters Limited (1983) *Science and Technology in Latin America*. Harlow: Essex: Longman.

Lawler, A. (1977) Academy seeks government help to fight openness law. *Science*, **277**, 473.

Lederman, L. (1987) Science and technology policies and priorities: a comparative analysis. *Science*, **237**.

Lipset, S. M. (1995) *American Exceptionalism: A Double-Edged Sword*. New York: W. W. Norton.

Lithwick, H. (1969) *Canada's Science Policy and the Economy*. Toronto: Methuen.

Lock, P. and Voss, W. (1997) Germany. In P. Gummett and J. Stein (eds), *European Defence Technology in Transition*. Amsterdam: Harwood.

Macdonald, S. and Reams, D. (1989) Selling government information: the issues overseas. Attachment B. In T. Madeville, *The Supply of Queensland Government Information*. Information Research Unit Department of Economics, University of Queensland.

Machlup, F. (1980) *Knowledge: Its Creation, Distribution, and Economic Significance. Vol 1: Knowledge and Knowledge Production*. Princeton, NJ: Princeton University Press.

Macintyre, S. (1994) *Whereas the People ... Civics and Citizen Education: Report of the Civics Expert Group*. Canberra: AGPS.

Maddison, A. (1991) *Dynamics forces in Capitalist development*. New York: Oxford University Press.

Madeville, T. (1989) *The Supply of Queensland Government Information*. Information Research Unit Department of Economics, University of Queensland.

Mankiw, N.-G. (1995) *Macroeconomics*. New York: Worth.

Mann, M. (1986) *The Sources of Social Power, Vol. 1: A History of Power from the Beginning to A.D. 1760*. Cambridge: Cambridge University Press.

Marceau, J. (1996) Rational expectations? Education policy and the language of management in Australia. In S. Clegg and G. Palmer (eds) *Constituting Management*. Berlin and New

York: W. de Gruyter.

Marceau, J. and Jureidini, R. (1992) Giants and Dwarves: changing technologies and productive interlinkages in Australian manufacturing industry. In J. Marceau (ed.) *Reworking the World: Organisations, Technologies and Cultures in Comparative Perspective.* Berlin: de Gruyter.

Marcuse, H. (1966) *One Dimensional Man.* Boston, MA: Beacon Press.

Margetts, H. (1991) The computerisation of social security: the way forward or a step backwards? *Public Administration,* **69,** 4.

Matthews, J. (1997) Power Shift. *Foreign Affairs,* **76.**

May, T. (1996) Defence contract row erupts as ITT allies with Racal. *Guardian,* 4 December.

Mayntz, R. (1996) Science in East Germany – Consequences of Unification. In W. Krull and F. Meyer-Krahme (eds) *Science and Technology in Germany,* new edn. London: Cartermill.

Mazur, A. (1974) Disputes between experts. *Minerva,* **11.**

Mazur, A. and Lee, J. (1993) Sounding the global alarm: environmental issues in the US national news. *Social Studies of Science,* **23.**

McCombie, C. (1997) Nuclear waste management worldwide. *Physics Today,* June, 56–62.

McDougall, W. A. (1985) *The Heavens and the Earth; A Political History of the Space Age.* New York: Basic Books.

McGeorge, B. (1989) *Danger and Survival: Choices About the Bomb in the First Fifty Years.* New York: Random House.

Medawar, P. (1967) *The Art of the Soluble.* London: Unwin Allen.

Medvedev, Z. A. (1969) *The Rise and Fall of T. D. Lysenko.* New York: Columbia University Press.

Melman, S. (1970) *Pentagon Capitalism: The Political Economy of War.* New York: McGraw-Hill.

Metcalfe, S. (1989) Trade, technology and evolutionary change. In R. Harrington *et al.* (eds) *Money, Trade and Payments.* Cambridge, MA: MIT Press.

Metcalfe, S. (1995) The economic foundations of technology policy. In P. Stoneman (ed.) *Handbook of the Economics of Innovation and Technical Change.* Oxford: Blackwell.

Metcalfe, S. and Soete, L. (1984) Notes on the evolution of technology and international competition. In M. Gibbons *et al.* (eds) *Science and Technology Policy in the 1980s and Beyond.* London: Longman.

Meyer-Krahmer, F. (1990) *Science and Technology in the Federal Republic of Germany.* London: Longman.

Meyers, S. and Schipper, L. (1992) World energy use in the 1970s and 1980s: exploring the changes. *Annual Review of Energy and the Environment,* **17.**

Meynaud, J. (1968) *Technocratie et politique.* Paris: Bellanger, La Ferbe-Bernard. Translated as *Technocracy.* New York: The Free Press.

Meynaud, J. and Schroeder, B. (1962) *Les Savants et la vie internationale.* Lausanne: Etudes de science politique.

Midttun, A. and Baumgartner, T. (1986) Negotiation energy futures: the politics of energy forecasting. *Energy Policy,* June.

Mills, S. (1986) *The New Machine Men: Polls and Persuasion in Australian Politics.* Victoria: Penguin.

Mitrany, D. (1966) *A Working Peace System.* Chicago: Quadrangle Books.

Mogee, M. E. (1973) *Public Policy and Organizational Change: The Creation of the RANN Program in the National Science Foundation.* M. S. Thesis, George Washington University.

Moorehead, A. (1952) *The Traitors.* New York: Scribner's.

Morin, A. J. (1993). *Science Policy and Politics.* Englewood Cliffs, NJ: Prentice-Hall.

Morrocco, J. (1977) EC outlines path for consolidation. *Aviation Week & Space Technology,* **147.**

Morton, L. (1969) War and social change. In K. H. Silvert (ed.) *The Social Reality of Scientific Truth*. Hanover, NH: American University Field Staff.

Mukerji, C. (1989) A fragile power: scientists and the state (Princeton: National Academy of Engineering, Committee on Public Engineering Policy. 1970). *Federal Support of Applied Research*. Washington, DC: National Academy of Engineering.

Nader, R. (1965) *Un-safe At Any Speed*. Washington, DC: Constitution Press.

National Science Foundation, National Patterns of R&D Resources: 1996, NSF 96–333. Arlington, VA: NSF.

Nelkin, D. (1972) *The University and Military Research: Moral Politics at MIT*. Ithaca, NY: Cornell University Press.

Nelkin, D. and Pollak, M. (1981) *The Atom Besieged: Extraparliamentary Dissent in France and Germany*. Cambridge, MA: MIT Press.

Nelson, R. R. (ed.) (1993) *National Systems of Innovation*. Oxford: Oxford University Press.

Nicholson, R., Cunningham, C. and Gummett, P. (1991) *Science and Technology in the UK*. London: Longman.

Nieburg, H. L. (1966) *In the Name of Science*. Chicago: Quadrangle Press.

Noam, E. (1992) *Telecommunications in Europe*. New York: Oxford University Press.

NSF (1997) *Annual Report*. Washington, DC: NSF.

Nunn, S. (1994) How capital technologies affect municipal service outcomes: the case of police mobile digital terminals and stolen vehicle recoveries. *Journal of Policy Analysis and Management*, **13**, 3.

Nye, J. S. Jr. (1990) *Bound to Lead: The Changing Nature of American Power*. New York: Basic Books.

Oden, M. (1996) Cashing-in, cashing-out and converting: restructuring of the defense industrial base in the 1990s. Paper presented to the Council of Foreign Relations Study Group on 'Consolidation, Downsizing and Conversion in the U.S. Military Industrial Base', New York, 9 February 1996 (mimeo).

O'Toole, K. (1994) GIS, administrative reform and local government: the issue of privacy. *Journal of Social Issues*, **29**, 4.

OECD (1971a) *Science, Growth and Society*, The Brooks Report.

OECD (1971b) *Information for a Changing Society: Some policy considerations*. Paris: OECD.

OECD (1994) *Science and Technology Policy: Review and Outlook, 1994*. Paris: OECD.

OECD (1996) *Science, Technology, and Industry Outlook, 1996*. Paris: OECD.

OECD (1999) *Main Science and Technology Indicators, 1999*, No. 1. Paris: OECD.

Osborne, D. and Gaebler, T. (1993) *Reinventing Government*. New York: Plume.

Parker, L. (1997) *The Engineering Research Centers (ERC) Program: An Assessment of Benefits and Outcomes*. Arlington, VA: National Science Foundation.

Patel, P. and Pavitt, K. (1995) Patterns of technological activity: their measurement and interpretation. In P. Stoneman (ed.) *Handbook of the Economics of Innovation and Technological Change*. Oxford, UK and Cambridge, MA: Blackwell.

Penrose, R. (1997) *Shadows of Consciousness*. Oxford: Oxford University Press.

Poguntke, (1993) *Alternative Politics: The German Green Party*. Edinburgh: Edinburgh University Press.

Polanyi, M. (1957) *The Tacit Dimension*. New York: Anchor Books.

Polanyi, M. (1962) *The Republic of Science: Its Political and Economic Theory*. Chicago: Roosevelt University.

Pollitt, C. (1996) Antistatist reforms and new administrative directions: public administration in the United Kingdom. *Public Administration Review*, **56**, 1.

Popper, S. W. (1991) *Science and Technology in Eastern Europe After the Flood: Rejoining the World*. Santa Monica, CA: RAND.

Posner, M. (1961) International trade and technological change. *Oxford Economic Paper*, **13**.

Price, D. K. (1954) *Government and Science; Their Dynamic Relation in American Democracy*. New York: New York University Press.

Price, D. K. (1965) *The Scientific Estate*. Cambridge, MA: Harvard University Press.

Price, D. K. (1984) The established dissenters: scientists and America's unwritten constitution. In V. Bogdanor (ed.) *Science and Politics: The Herbert Spencer Lectures 1982*. London: Clarendon Press.

Primack, J. and von Hippel, F. (1974) *Advice and Dissent: Scientists in the Political Arena*. New York: Basic Books.

Rahman, A. (1990) *Science and Technology in India, Pakistan, Bangladesh, and Sri Lanka*. Harlow, Essex: Longman.

Ravetz, J. R. (1971) *Scientific Knowledge and its Social Problems*. London: Oxford University Press.

Reid, R. W. (1969) *Tongues of Conscience: Weapons Research and the Scientists' Dilemma*. New York: Walker and Company.

Reingold, N. (1987) Vannevar Bush's new deal for research, or the triumph of the old order. *Historical Studies of the Physical and Biological Sciences*, **17**.

Reppy, J. (forthcoming) Back to the future: dual-use technology and proliferation. *Bulletin of the Atomic Scientists*.

Rettig, R. A. (1977) *Cancer Crusade: The Story of the National Cancer Act of 1971*. Princeton: Princeton University Press.

Rhodes, R. (1986) *The Making of the Atomic Bomb*. New York: Simon and Schuster.

Rhodes, R. (1995) *Dark Sun: The Making of the Hydrogen Bomb*. New York: Simon and Schuster.

Rhodes, R. (1996) The new governance: governing without government. *Political Studies*, **44**.

Rifkin, J. (1995) *The End of Work*. New York: Putnam.

Roberts, W. (1991) Geographic information systems: a local government perspective, *Memo*, 89.

Ronayne, J. and Boag, C. (1989) *Science and Technology in Australasia, Antarctica and the Pacific Islands*. London: Longman.

Rose, H. and Rose, S. (1970) *Science and Society*. Harmondsworth, Middlesex: Pelican Books.

Rostow, W. W. (1971) *Stages of Economic Growth*. New York: Cambridge University Press.

Rotblat, J. (1972) *Scientists in the Quest for Peace: A History of the Pugwash Conferences*. Cambridge, MA: MIT Press.

Rothenberg, J. (1992) Intertemporal and inter-generational anomalies for international response to global climate change. In N. Choucri (ed.), *Global Change: Environmental Challenges and International Responses*. Cambridge, MA: MIT Press.

Rouban, L. (1988) *L'État et la science: la politique publique de la science et de la technologie*. Paris Éditions du Centre National de Recherche Scientifique.

Rycroft, R. and Kash, D. (1999) Complex technology and community: implications for policy and social science. *Research Policy*, **23**.

Sabatier, P. (1988) An advocacy coalition framework of policy change and the role of policy-oriented learning therein. *Policy Sciences*, **21**.

Salamon, L. M. (1995) *Partners in Public Service: Government-Non-profit Relations in the Modern Welfare State*. Baltimore, MD: Johns Hopkins University Press.

Salomon, J.-J. (1970) *Science et Politique*. Paris: Editions du Seuil. Translated as *Science and Politics*. Cambridge, MA: MIT Press (1973).

Salomon, J.-J. (1993) *Le Destin technologique*. Paris: Balland.

Samuels, R. (1987) *The Business of the Japanese State*. Ithaca: Cornell University Press.

Samuels, R. (1994) *'Rich Nation Strong Army': National Security and the Technological Transformation of Japan*. Ithaca, NY: Cornell University Press.

Sandel, M. (1996) *Democracy's Discontent*. Cambridge, MA: Harvard University Press.

Sandholtz, W. (1992) *High-Tech Europe: The Politics of International Cooperation.* Berkeley: University of California Press.

Sapolsky, H. M. (1990) *Science and the Navy: the History of the Office of Naval Research.* Princeton: Princeton University Press.

Sardar, Z. (1982) *Science and Technology in the Middle East: A Guide to Issues, Organizations, and Institutions.* London: Longman.

Saxenian, A. (1994) *Regional Advantage: Culture and Competition in Silicon Valley and Route 128.* Cambridge, MA: Harvard University Press.

Scharf, T. (1994) *The German Greens: Challenging the Consensus.* Oxford: Berg Publishers.

Schelling, T. C. (1995) Intergenerational discounting. *Energy Policy,* **23** (4/5).

Scherer, F. M. (1996) *Industry Structure, Strategy and Public Policy.* New York: HarperCollins College Publishers.

Schroeder-Gudehus, B. (1973) Challenges to transnational loyalties: international organizations after the First World War. *Science Studies,* **3**.

Sclove, R. E. (1996) Town meetings on technology. *Technology Review,* **5**.

Seibel, W. (1996) Administrative science as reform: German public administration. *Public Administration Review,* **56**, 1.

Serfati, C. (1995) *Production d'Armes, Croissance et Innovation.* Paris: Economica.

Schibeci, I., Shaw, R. and Davison, A. (1999) Genetic medicine an experiment in community-expert interaction, *Journal of Medical Ethics,* **4**.

Shabecoff, P. (1993) *A Fierce Green Fire: The American Environmental Movement.* New York: Hill and Wang.

Shaiken, H. (1985) *Work Transformed and Labour in the Computer Age.* New York: Holt, Rinehart and Winston.

Shane, S. (1994) *Dismantling Utopia: How Information Ended the Soviet Union.* Chicago: Ivan Dee.

Shapira, P. (1997) Manufacturing extension: performance, challenges, and policy issues. In Branscomb (ed.) *Investing in Innovation.* Cambridge, MA: MIT Press.

Shapley, W. H. (1992) *The Budget Process and R&D.* New York: Carnegie Commission on Science, Technology, and Government.

Sherry, M. S. (1995) *In the Shadow of War: The United States Since the 1930s.* New Haven: Yale University Press.

Shilts, R. (1987) *And the Band Played On: Politics, People, and the AIDS Epidemic.* New York: St. Martin's.

Sigurdson, J. and Anderson, A. M. (1991) *Science and Technology in Japan,* 2nd edn. Detroit: Longman.

Skolnikoff, E. B. (1967) *Science, Technology and American Foreign Policy.* Cambridge, MA: MIT Press.

Skolnikoff, E. B. (1972) *The International Imperatives of Technology: Technological Development and the International Political System.* Berkeley, CA: Institute of International Studies, University of California.

Skolnikoff, E. B. (1985) Computers, armaments, and stability. In D. P. Donnelly (ed.) *The Computer Culture: A Symposium to Explore the Computer's Impact on Society.* Rutherford, NJ: Farleigh Dickinson University Press.

Skolnikoff, E. B. (1990) The Policy Gridlock on Global Warming. *Foreign Policy,* **79**, Summer.

Skolnikoff, E. B. (1993) *The Elusive Transformation Science, Technology and the Evolution of International Politics.* Princeton: Princeton University Press.

Skolnikoff, E. B. (1996) National innovation systems and global science and technology. Paper presented at Columbia University Conference, 20 September.

Skrabenek, P. (1990) Nonsense consensus. *Lancet,* 1446–7.

Slovic, P., Fischoff, B. and Lichtenstein, S. (1982) Rating the risks: the structure of expert and lay perceptions. In C. Hoehnemser and J. X. Kasperson (eds) *Risk in the Technological Society*. Boulder, CO: Westview Press.

Smith, A. K. (1965) *A Peril and Hope: the Scientists' Movement in America 1945–47*. Chicago: University of Chicago Press.

Smith, B. L. R. (1966) *The RAND Corporation: Case Study of a Nonprofit Advisory Corporation*. Cambridge, MA: Harvard University Press.

Smith, B. L. R. (1990) *American Science Policy Since World War II*. Washington, DC: Brookings Institution.

Smith, B. L. R. (1992) *The Advisers: Scientists in the Policy Press*. Washington: Brookings.

Smith, B. L. R. and Barfield, C. (1996) *Technology, R&D, and the Economy*. Washington, DC: The Brookings Institution and American Enterprise Institute.

Smith, J. (1994) *Managing Privacy: Information Technology and Corporate America*, Chapel Hill: The University of North Carolina Press.

Snellen, I., Van de Donk, W. and Baquaist, J. (eds) (1989) *Expert Systems in Public Administration: Evolving Practices and Norms*. Amsterdam: Elsevier Science Publishers.

Snow, C. P. (1981) *Science and Government*. Cambridge, MA: Harvard University Press.

Soete, L. and Verspagen, B. (1992) Competing for growth: the dynamics of technology gaps. Convergence and innovation. Paper presented at the Conference on Economic Growth and the Structure of Long-term Development, Varenna, 1–3 October.

Solingen, E. (ed.) (1994) *Scientists and the State: Domestic Structures and the International Context*. Ann Arbor: University of Michigan Press.

Sorenson, T. (1987/8) The President and the Secretary of State. *Foreign Affairs*, **66** (5), Winter 1987/8, 236.

Sproull, L. and Kiesler, S. (1991) *Connections: New Ways of Working in the Networked Organization*. Cambridge, MA: MIT Press.

Stace, D. and Norman, R. (1995) *Reinvented Government: The New Zealand Experience*. Centre for Corporate Change, Paper No. 050, Australian Graduate School of Management: University of New South Wales.

Stern, P. M. with Green, H. P. (1969) *The Oppenheimer Case: Security on Trial*. New York: Harper and Row.

Stewart, F. (1978) *Technology and Underdevelopment*. London: Macmillan.

Stockholm International Peace Research Institute (1997) *Sipri Yearbook 1997*. Oxford: Oxford University Press.

Strickland, S. P. (1972) *Politics, Science and Dread Disease: A Short History of U.S. Medical Research Policy*. Cambridge: Harvard University Press.

Strickland, S. P. (1989) *The Story of the NIH Grants Program*. Lanham, MD: University Press of America.

Studer, K. E. and Chubin, D. (1980) *The Cancer Mission: Social Contexts of Biomedical Research*. Beverly Hills, CA: Sage.

Synthesis Panel, The (1991) *Policy Implications of Greenhouse Warming*. Washington, DC: National Academy Press.

Tang, T. B. (1984) *Science and Technology in China*. London: Longman.

Teich, A. H. and Pace, J. H. with Adler, R. G. *et al.* (1986) *Science and Technology in the USA*. London: Longman.

Teich, A. H. (ed.) (1974) *Scientists and Public Affairs*. Cambridge, MA: MIT Press.

Teubal, M., Foray, D., Justman, M. and Zuscovitch, E. (eds) (1996) *Technological Infrastructure Policy: An International Perspective*. The Jerusalem Institute for Israel Studies. Dordrecht: Kluwer.

Thompson, E. P. (1966) *The Making of the English Working Class*. New York: Random House.

Tulder, R. van and Junne, G. (1988) *European Multinationals in Core Technologies*. London: John Wiley.

Tulder, R. van and Dankbarr, B. (1992) The illusion of a common supranational interest. In J. Marceau (ed.) *Reworking the World: Organisations, Technologies and Cultures in Comparative Perspective*. Berlin: de Gruyter.

Tyson, L. (1992) *Who's Bashing Whom? Trade Conflict in High Technology Industries*. Washington, DC: Institute for International Economics.

U.S. Senate (1971) *Conquest of Cancer Act, 1971*. Hearings before the Subcommittee on Health of the Committee on Labor and Public Welfare, U.S. Senate, Ninety-second Congress. Washington, DC: U.S. Government Printing Office. Quoted in Studer and Chubin, p.75.

Utterback, J. M. (1994) *Mastering the Dynamics of Innovation*. Boston: Harvard Business School Press.

Uyehara, C. H. (1966) Scientific advice and the Nuclear Test Ban Treaty. In Sanford Lakoff (ed.) *Knowledge and Power*. New York: Free Press.

Van Scherpenberg, J. (1997) Transatlantic competition and European defense industries: a new look at the trade-defense linkage. *International Affairs*, **73**.

Velocci, A. L. Jr. (1995) Profit wave uncovers nagging paradoxes. *Aviation Week & Space Technology*, **142**.

Velocci, A. L. Jr. (1996) Lockheed Martin targets major cuts. *Aviation Week & Space Technology*, **144**.

Vernon, R. (1966) International investment and international trade in product cycle. *Quarterly Journal of Economics*, **80**.

Verspagen, B. (1990) Catching up or falling behind? A dynamic model of growth rate differential. Merit Research Memorandum, 90-004, MERIT. Masstricht.

Verspagen, B. (1991) Technology, specialization patterns and growth rate differential: a multi-sector evolutionary model of balance of payments restricted growth. Merit research Memorandum, 91-007, MERIT, Masstricht.

Vig, N. (1968) *Science and Technology in British Politics*. Oxford: Pergamon Press.

Vig, N. J. (1992) Parliamentary technology assessment in Europe: comparative evolution. Paper presented to the annual meeting of the American Political Science Association, September.

Vig, N. and Kraft, M. E. (eds) (1994) *Environmental Policy in the 1990s*. Washington, DC: Congressional Quarterly Press.

Von Hippel, E. (1988) *The Sources of Innovation*. New York: Oxford University Press.

Von Hippel, F. (1991) *Citizen Scientist*. New York: American Institute of Physics.

Wade, R. (1990) *Governing the Market. Economic Theory and the Role of Government in East Asian Industrialization*. Princeton: Princeton University Press.

Walker, W. and Gummett, P. (1993) Nationalism, internationalism and the European defence market, Chaillot Papers, no. 9. Paris: Institute for Security Studies.

Wallace, D. (1995) *Environmental Policy and Industrial Innovation: Strategies in Europe, the U.S., and Japan*. London: Earthscan.

Wallerstein, M. B. (ed.) (1984) *Scientific and Technological Cooperation Among Industrialized Countries*. Washington, DC: National Academy Press.

Weaver, R. K. and Rockman, B. A. (1993) Assessing the effects of institutions. In Weaver and Rockman (eds) *Do Institutions Matter?* Washington: Brookings.

Weinberg, A. M. (1963) Criteria for scientific choice. *Minerva*, **1**.

Wildavsky, A. (1995) *But Is It True? A Citizen's Guide to Environmental Health and Safety Issues*. Cambridge, MA: Harvard University Press.

Williams, L. (1995) Computer gap worries blacks. *New York Times*, 25 May.

Wohlstetter, A. (1964) Strategy and the natural scientists. In R. Gilpin and C. Wright (eds),

Scientists and National Policy Making. New York: Columbia University Press.

Wong, P.-K. (1995) Computing in the global electronics industry: a comparative study of the innovation networks of Singapore and Taiwan. *Journal of Industry Studies*, **2**, 2.

Wood, R. C. (1964) Scientists and politics: the rise of an apolitical elite. In R. Gilpin and C. Wright (eds) *Scientists and National Policymaking.* New York: Columbia University Press.

Woolley, M. (1997) The comfort zone. *Science*, **275**, 1243.

York, H. F. (1970) *Race to Oblivion: A Participant's View of the Arms Race.* New York: Simon and Schuster.

York, H. F. (1976) *The Advisers: Oppenheimer, Teller, and the Super Bomb.* San Francisco: W. H. Freeman.

York, H. F. (1987) *Making Weapons, TalkingPeace.* New York: Basic Books.

Zacher, M. W. (1991) The decaying pillars of the Westphalian Temple: implications for international order and governance. In J. N. Rosenau (ed.), *Governance Without Government.* Cambridge: Cambridge University Press.

Ziegler, J. N. (1997) *Governing Ideas: Strategies for Innovation in France and Germany.* Ithaca: Cornell University Press.

Index

Page numbers in italics refer to Figures.

T - #0647 - 101024 - C0 - 244/170/14 - PB - 9780826450265 - Gloss Lamination